Asteroids in Synastry

Emma Belle Donath

Copyright 1978 by Emma Belle Donath
All rights reserved.

No part of this book may be reproduced or transmitted in any orm or by any means electronic or mechanical, including photocopying r recording. or by any information storage and retrieval system. without w en permission from the author and publisher. except in the case of brief quotations embodied in critical reviews and articles. Requests and inquir es may be mailed to: American Federation of Astrologers. Inc., 6535 S. Rural Road, Tempe, Arizona 85283.

First Printing: 1993
Current Printing: 2009

ISBN-10: 0-86690-082-9
ISBN-13: 978-0-86690-082-9

Cover Design: Jack Cipolla

Published by:
American Federation of Astrologers, Inc.
6535 S. Rural Road
Tempe, Arizona 85283

www.astrologers.com

Printed in the United States of America

Books by Emma Belle Donath

Asteroids in Synastry
Asteroids in Midpoints
Asteroids in the Birth Chart
Asteroids in the U.S.A.
Have We Met Before?
Houses: Which and When
Minor Aspects Between Natal Planets
Patterns of Professions
Relocation
Approximate Positions of Asteroids 1900-1999

The Planetoids

Life on the earth is but the reflection of a greater plan
The patterns of the cosmos—the action of woman and man.
Relationships not based upon the giants of life,
But upon the myriad of happenings creating balance or strife.
Moments that become the years,
The tiny breaths one hardly hears.
The stars reveal the story in a similar way.
The patterns of relationships day to day . . .
Found in the movement of the little planetoids,
Not in just what the giants say.
From Juno comes the message of wrong and right.
From Ceres comes service and nurturing delight.
Vesta calls for sacrifice and doing what's best.
Pallas Athena teaches building and defending the nest.
Their movements are subtle but surely felt
As we again seek the perfection when two become one.
The answers reflected through the asteroid belt.
Reveal the perfection Virgo would become . . .
The responsibility of home and the sanctity of life.
The joy of service and the bonds of man and wife.
Doing it right now so that we can enjoy it tomorrow.
To experience both the pleasure and the sorrow.
Myriam Ruthchild, 1977

Contents

Preface	ix
Chapter I, Asteroids and Relationships	1
Chapter II, Automatic Actions and Responses	15
Chapter III, Communications	19
Chapter IV, Companions	31
Chapter V, Comparison	39
Chapter VI, Completion	47
Chapter VII, Composite	59
Chapter VIII, Contact Cosmogram	69

Preface

Though individual asteroids or planetoids orbiting between Mars and Jupiter may be small, they are being found to exert influences in the horoscope. Several astrologers in the United States and Germany are presently experimenting with these effects in the natal chart, and by transit to and from the four major asteroids as well as their progressed and directed positions.

Theories concerning the formation of the asteroid belt vary from the possibility of a planet which exploded from internal or external pressure to the concept of two planets colliding in space or to the idea that gases present in that area simply did not adhere togethe into the dense mass necessary for a planet.

Since work with the asteroids has shown such concern with the field of human relationships, I began including them in synastry counseling. The additional information shown by adding these tiny bodies to horoscope is given in this volume.

It has been very helpful to be allowed to use data gathered from several large organizations and social groups in exchange for participating in their charitable endeavors. This has enabled me to include data obtained from happily married couples, harmonious partnerships, compatible groups and families as well as those problem cases which came to us for personal or compatibility counseling. Hopefully, this will be of aid to those astrologers who do premarital counseling.

A special thank-you to each of the following persons for help in preparing this volume: Lois Daton of Dayton, Ohio; Luisa Burns of Indianapolis, Indiana; Carol Jaramillo of Columbus, Ohio; Betty King of Miamisburg, Ohio; and Myriam Ruthchild of Cincinnati and Dayton, Ohio. Last, but certainly not least, my husband Robert C. Donath without whom I would never have learned to use my Juno in Leo properly.

Emma Belle Donath
Dayton, Ohio, 1977

Chapter 1

Asteroids and Relationships

Astronomical Data

Astrological information may be gleaned from an astronomical study of all heavenly bodies, including the little planets called asteroids. They were so named because they appear starlike when viewed through the average telescope. Their discovery was delayed because they are so faint that only Vesta can be seen with the unaided eye of man.

Table 1. Relative Brightness and Size of the Asteroids Compared with the Moon and Mercury

Object	Diameter (km)	Relative Brightness	Discovered
Ceres (1)	770	0.06	1801
Pallas (2)	490	0.07	1802
Juno (3)	193	0.12	1804
Vesta (4)	386	0.26	1807
Mercury	5000	0.069	-
Moon	3476	0.073	-

These relatively small particles were discovered at the beginning of the last century due to a discrepancy in Bode's Law. German scientist Johann Bode found that planets revolve around our sun in certain geometric relationships. According to his calculations there should have been a body between the orbits of Mars and Jupiter. An extension of this law also led to the discovery of Uranus and Neptune.

Table 2. Positions of the Planets in the Solar System According to the Law Devised by Bode

Planet	Distance from Sun to Earth (AU)	
	Predicted	Actual
Mercury	0.4	0.39
Venus	0.7	0.72
Earth	1.0	1.0
Mars	1.6	1.52
-	2.8	-
Jupiter	5.2	5.2
Saturn	10.0	9.5
Uranus	19.6	19.2

During the early 1800s four of the asteroids—Ceres, Pallas Athena, Juno and Vesta—were sighted by leading astronomers and their positions were added in 1870 to the yearly almanacs. By 1900 approximately 400 bodies had been sighted in the asteroid belt.

Even though the observation of the first four asteroids took several years, today there have been sightings of more than 5,000, most of which are numbered rather than named. These first four—Ceres, Pallas Athena Juno and Vesta—are the ones with which we are presently concerned because we have more data on their movements and eccentricities.

French astronomer Urbain Jean Joseph Leverrier, who discovered the planet Neptune in 1846 by calculations based on orbital irregu-

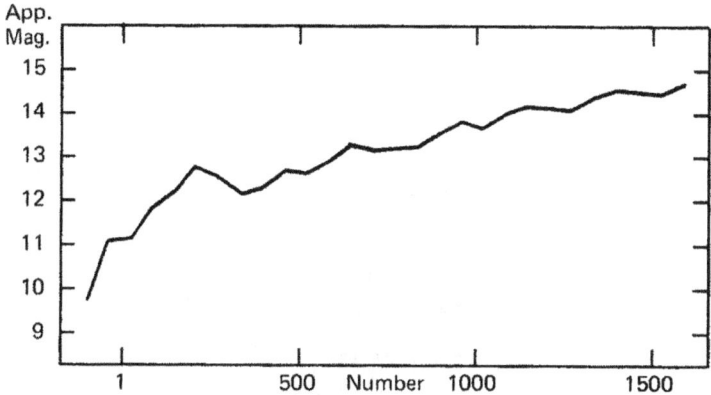

Figure 1. Asteroids Were Discovered According to Their Relative Brightness as Seen from Earth

larities of Uranus, included both the asteroid belt and Vulcan in his positioning of the planets. The asteroids are shown as "Petites Planets" on his commemorative medal.

Some data are available concerning other of the larger steroids. Before World War II the Astronomische Rechen-Institute at Dahlem, near Berlin, was the world center for asteroid research, even to the point of publishing a yearly ephemeris of the larger bodies. Other planetoids in order of their date of discovery include: Astraea (1845), Hebe (1847), Iris (1847), Flora (1847), Metis (1848) and Hygeia (1849). Certain bodies are important because of their eccentric orbits such as Hidalgo, Gounessia, Lipperta, Amor, Apollo, Adonis and Hermes.

It is unfortunate that many of the names are Greek or Roman versions of previously named planets. Or is this a suggestion to carefully research the finite connotations between Roman and Greek versions of any word? Even though the major characteristics of the gods remained the same when translated into another tongue, sometimes observation of ritual or sacrifice differed slightly; yet

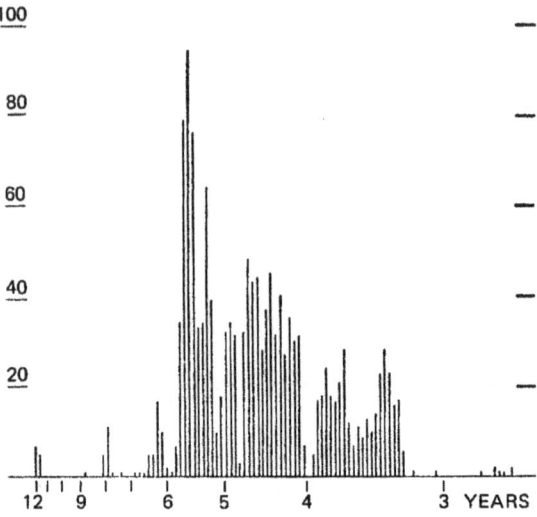

Figure 2. Orbital Cycles of the Asteroids

other times scant changes were made in the appearance or clothing of said deity. Perhaps the time has arrived for precise differentiation and discrimination.

The typical asteroid orbit is more elongated than most planetary orbits. Most of them require between 3.5 and 6 years to orbit around the Sun or have a 3 to 6 year cycle. There is a tendency for these bodies to travel parallel to Jupiter's orbit. The full extent of Jupiter's influence is demonstrated by the ratio of their gaps to Jupiter's 11.9 year orbit. The most conspicuous gaps occur at 5.95, 4.76 and 3.97 years, or exactly one-half, two-fifths and one-third respectively. This would suggest a harmonic relationship between the families of asteroids.

At first astronomers found the asteroids clumped into five loosely associated groups or families. More recent investigation has shown at least seven so-called asteroid families, maybe as many as 24. Use of these clumpings will probably be found to give more ac-

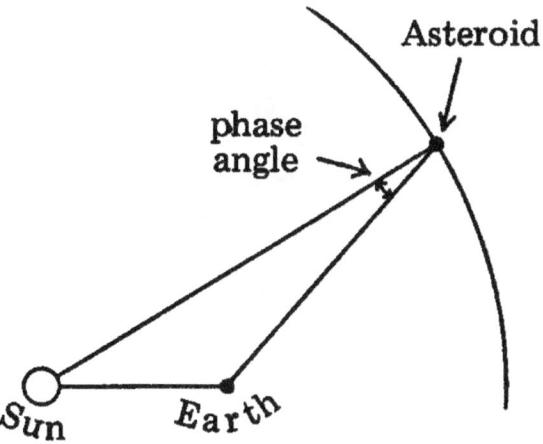

Figure 3. Angle of Observation of the Asteroids from the Earth

curate astrological timing than the precise locations of any single asteroid. Only time and extensive research can tell. So far it has been possible to time events with Ceres and Pallas on the Uranian 90-degree wheel using available positions. For most natal work, however, with orbs of up to five degrees, such precision is not ordinarily used. Some of my coworkers have suggested that even the computer positions used for the four major asteroids may vary by as much as 10 degrees from their actual positions at different periods. It will be interesting to compare the projected positions with astronomical observations in future years.

Only a few asteroids orbit near the plane of the earth's motion, but many are tilted at sizeable angles from the ecliptic. For example, Pallas Athena orbits at a variance of over 34 degrees from the ecliptic. This will be interesting to observe in delineation of declinations. Despite these differences, the minor planetoids do orbit in the same direction as the major planets, or astrologically they are direct in motion except for their regular retrograde periods.

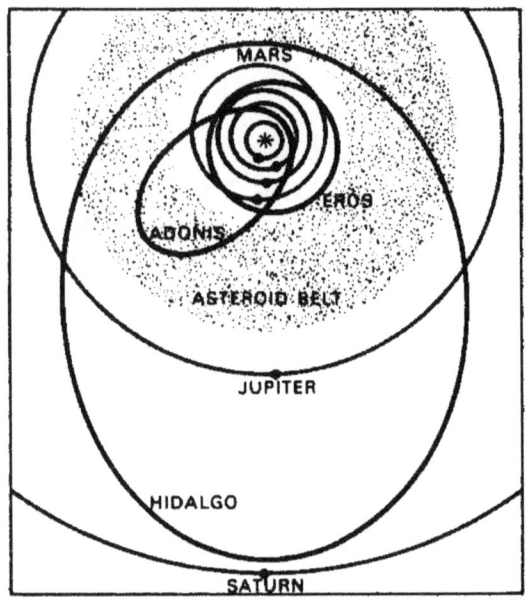

Figure 4. Asteroid orbits showing one with great variance. Orbits of typical asteroids are tilted 10 degrees. Hidalgo is tilted 42.5 degrees, even more than Pallas.

Since asteroids travel in clusters or groups, it stands to reason that they *deal with relationships from the interpretive standpoint.* Bodies which do not adjust and blend into the movements of others crash and break apart. Cooperating bodies continue to speed through their heavenly pathways undisturbed.

Another interesting reference was found concerning the asteroids and relationships. The concluding remark in the *Encyclopedia Americana* section on asteroids states: "International cooperation has been necessary for both the observation of asteroids and computation of their orbits." Even countries had to cooperate to find out about these little bodies. All of this gives clues to their astrological meanings as well.

To return to references about the approximate 4 to 6 year cycles, we would expect to begin to observe cyclic events in areas ruled or connected with the asteroids such as a 4 to 6 year crop variance, birth rate change, medical personnel availability, labor-management decisions, wedding-and-engagement ratio, and others. German astrologer Dr. Parm has found some interesting cyclic relations between Vesta and Saturn.

Asteroid returns, or 4 to 6 year cycles may be related to our school divisions. Many systems are divided into either 6 to 4 or 4-4-4, considering elementary, middle and high school. Most college degrees are given for either 4- or 6-year periods of study. Pallas has been found to be concerned with vocational or career training. Of course the suspected affiliation with the sign Virgo would be relevant here if Virgo does truly rule the school years between the early training of Gemini and the college studies of Sagittarius.

For further speculation on the possible leanings of these little bodies between Mars and Jupiter, look at the midpoints. In addition to my interpretation of the Mars/Jupiter midpoint as being potentially explosive, there are the following delineations in Witte's *Rules for Planetary Pictures* and Ebertin's *Combination of Stellar Influences,* respectively.

> "Mars/Jupiter—Joyous happenings, Betrothel, Fortunate deeds, To create something, To produce, Propagation, Pregnancy, Generation, Children, Fruits, Birth."

> "Mars/Jupiter—Agreements, Contracts, Getting engaged, Marriage, Births, Settlement of conflicts, Marital differences, Rebelliousness, Immoderation."

So far, these planets midway between Mars and Jupiter seem to prove that they not only serve as a buffer zone between the Martian energies and the Jovian expansiveness, but they also exemplify he midpoint delineations researched to date.

New information and tools are constantly being made available to encourage further research on the use of these four and other asteroids in astrological charts of all types. It is only by sharing the work done by individuals and groups that the importance or relevance of these bodies can be determined by astrologers of the present and future. To paraphrase Shakespeare, "It is better to have tried and lost than never to have tried at all."

Ancient Myths
Planets and stellar bodies are traditionally named for ancient Greek and Roman gods and goddesses. Strangely enough, the names given by astronomers seem to describe quite clearly the manifestations of these bodies in astrology. When the asteroids were discovered, astronomers began naming them for Roman goddesses.

Ceres, first to be tracked through the sky, was named for the Roman goddess of the harvest. This goddess who was worshiped under the name of Demeter in Greece, was a sister of Jupiter and one of the greatest of the deities. She was called "grain mother" or "earth mother" and ruled harvests, grains, the seasons, and fruitfulness. She was always depicted as a lovely maiden carrying a sheaf of wheat in her right arm. April 19 was observed as her feast day.

In earlier Etruscan times the equivalent goddess was shown carrying the sheaf of wheat, but in her left hand she held a sacrificial knife. She was said to rule underground waters and caves in addition to the harvest.

On of the most important goddesses of ancient Greece was named Athena or Pallas Athena, goddess of wisdom. When the second asteroid was named for this body, it too was found to indicate some of the characteristics of wisdom and natural intuition in the horoscope.

Pallas, favorite daughter of Jupiter, was supposed to have sprung full grown from his brow or be his "brain child." Like her father she wore the magic breastplate of goatskin fringed with snakes which produced thunderbolts when she was angered.

Even though considered a war goddess, she was not so much a fighter as a wise and prudent counselor. Pallas was called the defender of all cities and states.

Her symbol of the olive tree comes from the legend of the struggle between herself and Poseidon to gain the city of Athens. Jupiter had offered it to the god who presented the most useful gift to the people. Poseidon gave the horse, but Pallas' gift was the olive tree which could be used for healing, food and oil. She won the city of Athens.

Juno, third of the asteroids to be named, represented the wife of Jupiter and Queen of the Heavens. She was considered the goddess of women and marriage and the first day of each month was sacred to her.

In addition, Juno was worshiped at a shrine called Moneta from the Latin word counsel. From this term came the words "money" and "mint." Money was minted in a building attached to the temple.

Juno and Jupiter did not lead a harmonious life because of Jupiter's infidelities and Juno's jealous temper. She was usually shown as a majestic woman of mature age, with large open eyes and pale white skin, having near her the peacock, cuckoo and pomegranate.

When the fourth asteroid was discovered, it was given the name Vesta for goddess of the hearth. Roman families worshiped Vesta as guardian of family life and keeper of the sacred fire.

Th perpetual flames in the temples were guarded by maidens or Vestal Virgins. They spent 30 vestal years in service—10 learning their duties, 10 performing their services, and 10 training new virgins into the temple rituals. These virgins were bound by vow to protect the sacred fire. Breaking of this vow for any reason resulted in being buried alive or stoned to death. Further information and myths are given in *Asteroids in the Birth Chart* .

The keywords on the following pages were derived from these mythological descriptions and later proven to function in horoscopes.

⚳

Ceres, Goddess of the Harvest

Animal: Dolphin, pigs

Bird: Dove, crane

Insect: Ant

Plant: Cereal grains, corn, poppies

Emblem: Scythe, torch, sacrificial knife

Principle: Nurturing, healing

Expression: Cultivation, harvest, domestication, nourishment, useful labor, austere beauty, instinctive racial feelings, heritage of primitive experience, initiator of future life, server, concern

Manifestation: Ecology, cereal grains, farm tools, toilet training, domestic animals and pets, harvest fairs, honey, bread, caves, underground water or springs

Personification: British nanny, servant, nurse, nursery school director, earth science teacher, farmer, salutary laws, grandparent, grandchild

Pallas Athena, Goddess of Wisdom

Animal: Unicorn

Bird: Owl, cock

Insect: Beetle or scarab

Plant: Olive tree, iris

Emblem: Diamond or shield, lance

Principle: Prudent intelligence, intuition

Expression: Conservation, personification of light, invention, modesty, peacemaking, bravery, liaison, insensitive to emotions, noble mind, perseverance, equality in work, cunning, skillful hands, valor, perception

Manifestation: Embroidery, weaving, job, flute, battle armor, industry, numbers, spinning, handcrafts, fulcrum point

Personification: Weaver, warrior for peace, crusader, patron of the arts, working woman, counselor, horse trainer, sculptor, librarian, vocational teacher, National Guards reservist, ombudsman

Juno, Goddess of Marriage

Animal: Cow

Bird: Peacock, cuckoo

Insect: Bee

Plant: Pomegranate, rose, lily

Emblem: Linked double rings, sceptre

Principle: Legal mating

Expression: Jealousy, feminine wiles, inequality in partnership or work, social patterns, domestic organization, fidelity, virtue, fruitfulness, subtlety, vindictiveness, eccentricity, grace, mating, shrewishness, birth, quarrelsome

Manifestation: Ceremonies, weddings, ornamentation, etiquette, conjugal honor, protocol, mint, cosmetics, jewelry

Personification: Bride, groom, co-ruler, co-leader, vice president, widow, hostess, housekeeper, financial guardian, wife or husband, matron, yokemaker, egg, sibling

Vesta, Goddess of the Hearth

Animal: Deer, donkey

Bird: Nightingale, ibis

Insect: Spider

Plant: Laurel, oak tree

Emblem: Tripod, flaming altar

Principle: Dedication, sacrifice

Expression: Purity, service, sacrifice, solemnity, chastity, virginity, barrenness, purification, preserver, security, period of servitude, zealousness, delay, warmth

Manifestation: Ceramics, rituals, sanctuary, hearth, fire, breaking bread together, care of ancestors, altars, lamps, houses

Personification: Keeper of traditions, priest's assistant, caretaker, nun or monk, acolyte, Altar Guild, club member, Mason, Rosicrucian

Chapter JJ

Automatic Actions and Responses

When a client requests a comparison of his or her chart with that of another, my first question is: "To what purpose?" Is this really a quest for further understanding of self and others, or merely curiosity about another? Synastry, or the astrology of relationships, can lead to a deeper understanding of self and a beginning of comprehension of some of the subconscious reactions of others. However, no person should ever invade another's privacy by hearing a horoscope interpretation without the native's permission.

If two people are in the early stages of acquaintanceship, a simple comparison or completion type of analysis suffices to determine what talents and expressions they bring into each others' lives. There too are found some of the fundamental problems they will cause each other to face. This is as valid for employer-employee relationships or informal group activities as for romantic situations.

When the persons involved are family related, such as parent-child, husband-wife or siblings, composite charts, not only of their two natal horoscopes but charts made to include other family members, are helpful. Such dates as the birth of a child, parental death or relocation, and marriage or death of relatives add pertinent information. In other words, the flow of life together can be seen by erecting composite charts for meaningful periods or changes in their lives. Although each individual will have personal feelings and responses according to his or her own horoscope, two or more persons bound together will form a bond that can be analyzed as though it were an actual entity. This is known as the composite chart. There are special techniques for progressing and/or directing this type of chart which are fully described in *Planets in Composite* by Robert Hand.

Composite charts are also useful in considering group situations. When one or more persons come together for periods of time to work intensively toward a common goal such as a play, class, committee, orchestra, band or sports team, the director, coach or teacher can make good use of a composite of the entire group by knowing how they will respond to instructions. Timing of this type of composite is helpful in setting up meetings, game times, opening nights, special trips or examinations. Groups of up to five persons may also be considered on the contact cosmogram, which is another good timing device.

It is in the family situation that karmic interpretation is most helpful. When clients are open to this type of esoteric information, karmic delineation may be added at the conclusion of the interview. If mention of reincarnation would upset or confuse the clients, such data may simply be considered as a subconscious reaction during the entire counseling session.

When timing very specific events in the lives of two persons, be they married couples, children or business partners, the Mercury Method of chart comparison (see Chapter III) is excellent.

Mercury Method uses minor aspects along with the major ones in small orbs related to the planet Mercury which itself moves rapidly. Therefore, by mathematical calculations, timing may be established.

Mercury Method also presents a thorough analysis of the manner in which two people communicate. It does not deal extensively with sexual or physical compatibility but enters the larger scope of general adaptability. Very few cases of genuine biological incompatibility are found in marital counseling, but these few do require a medical approach and solution. Therefore, it is always wise to look into the individual charts for these factors.

Synastry is both the conscious and subconscious reaction of two or more persons to each other in any given situation. Knowledge of these reactions, whether they be automatic or deliberate, should help each person to become more responsible for his or her daily responses and actions. As the Native American philosophers used to say, "Do not criticize until you have walked for a mile in another's moccasins." Seeing another person's inner feelings and motivations through the horoscope gives personal empathy to understand how each affects the other.

Chapter III

Communication

The Mercury Method of Chart Comparison by Lois Rodden introduced a completely new concept for considering compatibility. All relationship situations tend to narrow to either the ease or difficulty of communication. So what better way to evaluate the possibility of rapport than to look at aspects related to Mercury. Use this simple process:

1. Aspect Mercury from Chart A to every planet in Chart B.

2. Then aspect B's Mercury to every planet in A's chart. This should include the respective Midheavens and Ascendants.

Further information and details may be found by reading the above-mentioned book as well as *Synastry* by Robert Jansky.

This method is particularly effective when timing activity between two persons. Events can be dated by using the progressed aspects in exact culmination.

1. Progress both charts to the same time period.

2. Then aspect Chart A's progressed Mercury to all progressed planets and all natal planets from Chart B. Use only a one-degree orb to determine events for a given year.

3. Aspect Chart B's progressed Mercury to all natal and progressed planets from Chart A using the same one-degree orb.

These aspects will show events occurring between or to the persons involved during the time period chosen. Accuracy in timing is, of course, depends upon the accuracy of the total horoscope.

Table 3. Aspects and Keywords for Asteroids in Mercury Method of Chart Comparison

Degree	Aspect	Orb	Keywords
0	Conjunction	5	Identifies, shares
30	Semisextile	2	Enthusiasm
45	Semisquare	2	Critical
60	Sextile	3	Benefits
90	Square	5	Action, strife
120	Trine	5	Ease
150	Quincunx	3	Immediacy
180	Opposition	5	Contrast, separative

Planetary principles between Mercury and the asteroids Ceres, Pallas Athena, Juno and Vesta manifest as follows:

Mercury-Ceres
Since Ceres represents a nurturing principle, this indicates assistance or interference in all Mercury-related factors depending upon the respective aspects. The Ceres individual might be supportive of Mercury during school years—an excellent placement for good aspects between pre-school or elementary teacher and student. Though not necessary to cement relationships, Ceres brings cooperative efforts such as the wife who, being concerned about her husband's diet and physical activity, quietly manages the home, thus allowing his energies to be directed toward career advancement.

With negative aspects between Mercury and Ceres, the daily regimen could be so neglected or abused as to create nervous tension for the Mercury person—not an aggressive relationship even under square aspects.

Mercury-Pallas Athena
With both Pallas Athena and Mercury associated with mental concerns, they would be expected to be compatible regardless of aspect. However, when found in square aspect, Pallas demands discipline from the Mercury native, thus making Mercury feel restricted. Since Pallas is concerned also with controlling or taming energy aimed toward productivity, this individual will be irritated by Mercury's seeming haphazard daydreaming and scattering of energies.

In harmonious aspect, Mercury inspires Pallas into experimenting with new methodology. Even when found in opposition, these two can balance the energies by letting the Mercury individual contribute ideas with Pallas channeling these concepts into practical application. Such an activity as creative writing could be accomplished together.

In employer-employee relationships it works better for Pallas to be the manager or boss.

Mercury-Juno
Although we use Juno as the wife principle, the tendency is to forget that she was also Jupiter's sister, thus giving her special affinity with Mercury, the sibling. Like many sisters, there is often sibling rivalry and bickering manifested between these two planets, but there is always deep affection. There may be nagging and nitpicking where Juno and Mercury are found in negative aspects but also a certain protectiveness.

Juno often is severe and restrictive because of its own inner insecurities. When the Mercury individual follows his or her natural inclination to be pleasant and cheerful, he or she can make Juno more at ease. Where Mercury acts in a feisty or peevish manner, Juno assumes the strict parental role.

Quincunx aspects between these two planets show the need for revision of approach toward each other. In marital relationships the Juno person tends to treat the Mercury native as a child in matters concerning finances. This is better handled by revising the rebellion and demands into balanced cooperation.

Mercury-Vesta
To combine extreme zealousness with the gift of gab can be either stimulating or explosive. When Mercury and Vesta individuals are both convinced of the importance or value of a particular project, nothing can beat this combination for promoting and carrying out said project. However, if they do not agree precisely, Mercury will seem to frustrate Vesta by bombarding this quiet soul with unceasing words like the drops in water torture. It will really be the Mercury person who is suffering most from the silent treatment handed out by stoic Vesta.

It is easier for these two persons to relate on a mental rather than emotional level, especially if they are bound by tension aspects.

In daily living Mercury is irritated by Vesta's dedication. For example, at times when the Mercury native impulsively decides he or she is ready for fun and games, he or she is terribly upset by Vesta's adherence to the scheduled job or activity. But Mercury in trine or conjunction admires Vesta's dedication.

On the following pages a complete delineation using the Mercury Method is given for Charts 1 and 2, respectively as A and B.

With a good aspect between the two Mercury planets, there is an opportunity to develop lines of communication and understanding. These people are easily drawn together at one time during their lives for mutual interchange. Both must make an effort for a permanent, close association to develop. This was a second marriage for both parties and children from previous alliances are involved.

Mercury Aspects, Chart 1

Female	*Aspects*	*Male*
Natal Mercury	Sextile	Mercury
1° Leo	Square	Sun
	Square	Moon
	Square	Venus
	Trine	Mars
	Sextile	Juno
	Opposition	Pallas
	Opposition	Jupiter
	Trine	Saturn
	Square	Uranus
	Conjunction	Pluto
	Square	Ascendant

In addition to an exact square of their Suns, there is a square from Mercury to Sun. The female is drawn to the authority and strength of the male but he must show deferential respect to gain the male's full cooperation. If she wishes to be willful there will be a struggle for power. With much effort this aspect can mean a mature and significant relationship. They are still trying to work out the power struggle.

Though the female is drawn to the charm of the male, there are obstacles to a free expression of love in these charts. Without any planets in air signs, these two people must determine other means to express themselves. If the female remains cheerful there will be harmony in the relationship. They are involved in meditation and extrasensory perception techniques to alleviate this problem.

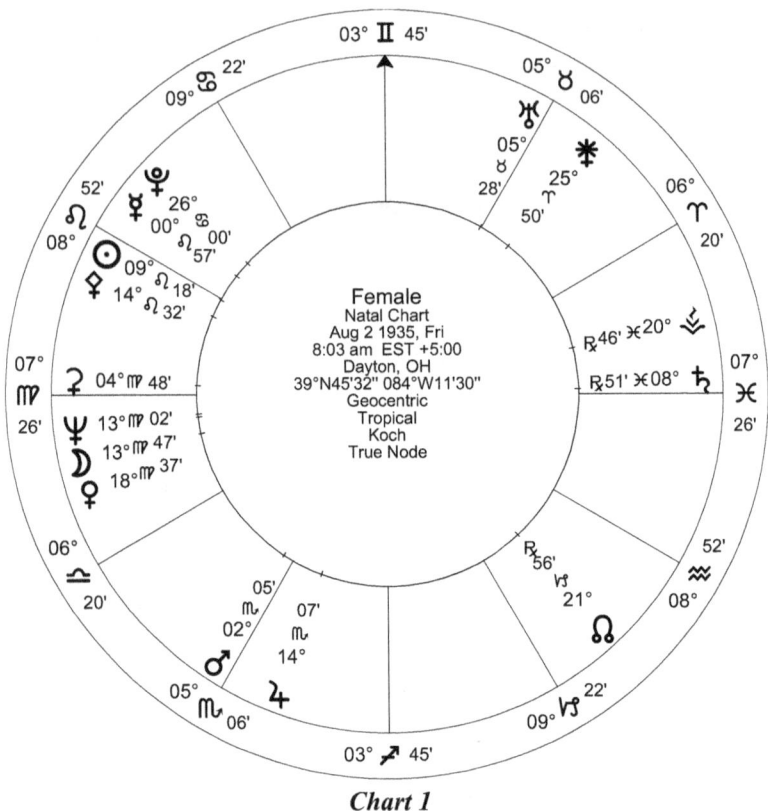

Chart 1

Mercury in square aspect to the Moon shows a female who is sympathetic to the male's moods and concerned for his welfare. If temperament is involved the contact could become tempestuous. There needs to be an equal exchange of nourishment and care, trust and tenderness here.

A courageous and honest female will get the full benefit of the constructive relationship promised by Mercury trine Mars. There will be periods of tension and excitement, vigor and sexual dynamics.

There is a great bond in the religious and philosophical understanding for these two people. Remember that each will grow in

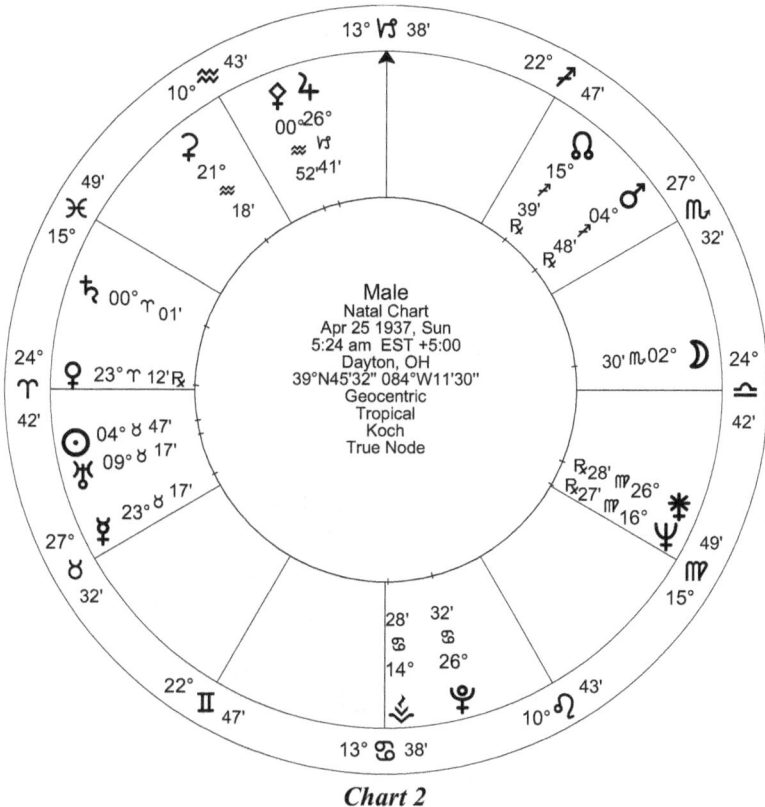

Chart 2

his or her own direction an must have freedom as well as loyalty, faith and trust. They are involved in spiritual study and work together, thus fulfilling the prior premise; but they have yet to learn to give each other the necessary freedom to develop.

An understanding female can draw upon the great strength from this male for the mutual benefit of all. If she acts in a self-centered way, she will take advantage of him. She must give him full credit for his abilities.

Mercury Aspects for Chart 2

Male	Aspects	Female
Natal Mercury	Sextile	Mercury
23° Taurus	Trine	Moon
	Trine	Venus
	Semisextile	Juno
	Sextile	Vesta
	Opposition	Jupiter
	Sextile	Pluto
	Conjunction	Midheaven

The independence and individuality of both are strong attractions for two strong people, but each must remember to build on the new and unusual ideas of the other while accepting the sudden and drastic changes they bring into each life. Eccentricity will cause an erratic relationship.

There may be periods when the female worries about not understanding this male. Criticism or irritation will only add to the confusion. Neptune aspects may be used in the higher vibration of sharing dreams and ideals.

Both identify with each other's higher dreams and aspirations and come together for the mutual benefit of groups and contacts. They have formed a non-profit institute for spiritual study. This Mercury conjunct Pluto aspect implies times of intensity as well.

Each of these persons has a different emotional pattern and they must respect the right of the other to express with separate friends and associates, which is hard for them to accept. Freedom is necessary for positive and direct work toward the common effort.

For the male there is a spontaneous admiration of the female. He can love her as a companion, friend or mate. If things are not harmonious he will tend to withdraw emotionally. With Mercury sextile Pluto the male again understands the dreams and ideals of the female and can tap the heights and depth of emotion with her. Timing is important between these two people, although they will cooperate for inner development of each other regardless of personality problems.

There is a certain level of aggression in each of these people which is incomprehensible to the other. This can react as passion or trauma.

Each chart shows the chance to express as the ideal mate for the other using the asteroid Juno as mate while Venus implies the type of women he prefers and Mars the men she would like best. These planets are shown in good or harmonious aspect to each other.

They will wish to begin a business together, but his Pallas opposition her Mercury makes him react in a stern and restrictive manner to her ideas for expansion. A struggle for power is again indicated.

He respects her dedication to her work with his Mercury sextile her Vesta. They both benefit from any undertakings about which they are sincere and dedicated.

This couple has now been married for two years and the reading is manifesting as suggested.

The second set of example charts have been progressed to 1974 to show the general principle of progression with this method. During this year the female's Mercury made a trine to the male's natal Juno and a quincunx to his progressed Juno, indicating marriage complications and involvements. He was in the process of divorcing his wife and preparing for remarriage during that year, and there were minor complications and irritations involved. Her Mercury conjunction his progressed Pallas shows the business venture they began jointly.

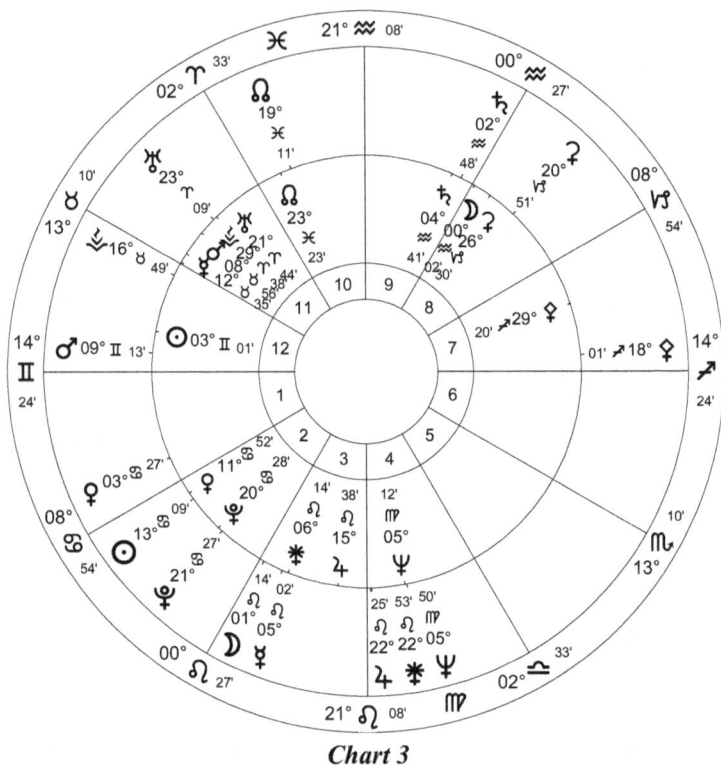

Chart 3

The male's Mercury, progressed to 1 Cancer, makes an opposition aspect to her natal Pallas, again indicating their combined business concern. His Mercury is moving from a sextile to her natal Vesta, indicating that he recently assisted her through a period of sacrifices and difficulties. Note the approaching aspect of the female's Mercury to the male's Ceres. In mid-1976 she nursed him through a hospitalization and convalescence.

Inner Wheel
Male
Natal Chart
Jun 4 1935, Tue
11:51 pm EST +5:00
Dayton, OH
39°N45'32" 084°W11'30"
Geocentric
Tropical
Koch
True Node

Outer Wheel
Male
Sec.Prog. SA in Long
Jun 4 1974, Tue
6:36:20 am EST +5:00
Dayton, OH
39°N45'32" 084°W11'30"
Geocentric
Tropical
Koch
True Node

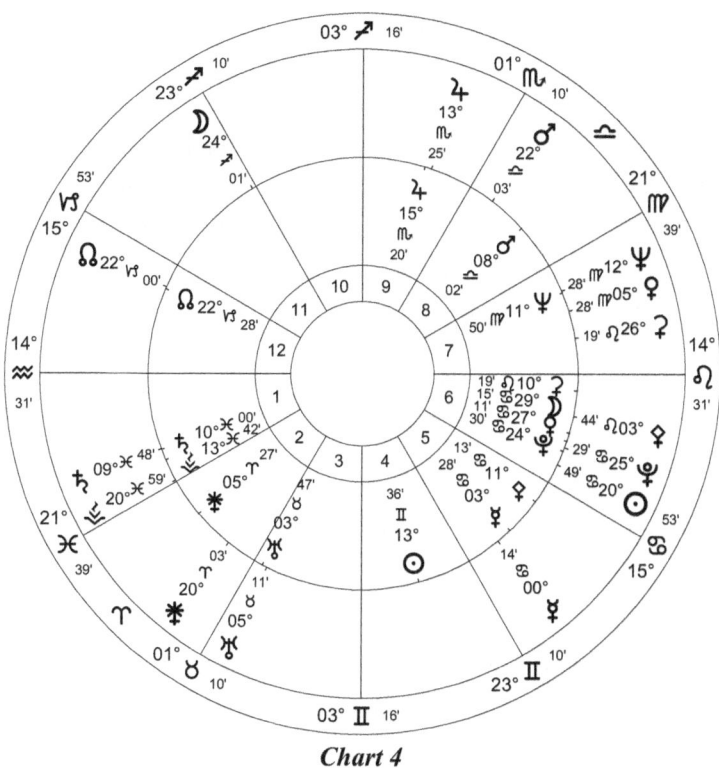

Chart 4

Charts 3 and 4 are shown with only the asteroids progressed to the given year with only the Mercury-asteroids aspects given on the aspect pages to highlight asteroid delineation. Further information can always be obtained by progressing both charts and aspecting all the planets from each chart, using both natal and progressed data. The Mercury aspects for both charts are shown on the next page.

Mercury Aspects for Chart 3

Female	*Aspects*	*Male*
Progressed Mercury	Trine	Natal Juno
	Sesquiquadrate	Progressed Juno
6° Leo	Conjunction	Progressed Pallas

Mercury Aspects for Chart 4

Male	Aspects	Female
Progressed Mercury	Opposition	Natal Pallas
	Sextile	Natal Vesta
1° Cancer		(Past 1 year)

Chapter IV

Companions

In working with four of the asteroids—Ceres, Pallas Athena, Juno and Vesta—there is a concern with relationships. Aspects to one or more asteroids have been found to one or more asteroids in both the coming into or the going out of relationships. One asteroid in particular is concerned with legalized relationships. It is this planetoid, Juno, which represents the legal mate in the horoscope.

The ancient Roman term for yokemaker became also the name for their married Goddess—Queen. Ceremonial bonds brought the yoke of responsibility. Juno, wife of Jupiter or Zeus, was the only married goddess among the Roman or Greek deities. Though she was known as Queen of the Sky she was in no way equal to Jupiter; nor could she control his unfaithfulness. This inequality represents the employee-employer relationship of legalized marriage up to our present age.

Juno was considered as woman deified. She was goddess not only of weddings but of matrons, child bearing, widows, guardian of national finances, and co-leader of the country. This was also the goddess of grace and eternal beauty. Juno was normally pictured as a beautiful woman, majestically gowned and bejeweled, carrying a golden sceptre. She was often surrounded by roses and lilies as well as strutting peacocks.

By legalizing relationships, Juno places them into established forms. Thus, the persons involved have publicly committed themselves to certain obligations and responsibilities. Wherever Juno is found in the horoscope, it tends to structure or formalize activities governed by that sign and house.

Juno in the radix describes how the native expects and wishes his or her mate to behave. The relationship that natal Mars or Venus have wit h natal Juno indicates whether this native will choose a mate living up to his or her expectations or not. For example, in a man's chart Venus in compatible aspect to Juno will show that he chooses a mate or mates who have similar qualities with those which he enjoys in a woman. For a woman's chart, the aspect from Mars to Juno will determine the desirability of trait in her chosen husband.

In any concern with synastry the chart of individual involved must be thoroughly studied discover how he or she is capable of handling any relationship. Quite often the problem in a marriage will be found in the horoscope of either the husband or wife when the comparative aspects all indicate harmony.

Following are some of the particular problems which can be considered with Juno in the native's own chart. Squares or oppositions are the two most prominent aspects, although this in no way eliminates other aspects for research or consideration. These are factors which can cause disturbances in a female's chart regardless of the man she marries:

Moon Square Juno in Female's Chart: Continuing irritation and problems from her mother about the marriage. Rather critic al of the husband about small details.

Moon Opposition Juno in Female's Chart: Torn between duty to mother or husband. Periodic separation from the spouse because of parental demands or needs such as prolonged illness.

Sun Square Juno in Female's Chart: Conflict between her father and husband. Possibility of an overactive Oedipus complex.

Sun Opposition Juno in Female's Chart: Lack of communication or common interests for father and spouse who would have quite different goals and philosophies. Disinterested in husband's career and advancement.

Mars Square Juno in Female's Chart: A feeling of being confined by the act of marriage. Makes better mistress than wife. Might be considered a female chauvinist.

Mars Opposition Juno in Female's Chart: Unable to understand husband's concern. Needs to earn to compromise rather than fight for own way in all types of relationships.

In a male's natal horoscope the following aspects may materialize regardless of his wife:

Moon Square Juno in a Male's Chart: Usually found in the chart of an only or youngest son. Close relationship with mother causes conflict because of jealousy mother has for mate. Must at some time break the apron strings.

Moon Opposition Juno in Male's Chart: Emotional coldness because of never feeling maternal love. Needs to learn how to express affection through real concern and warmth.

Sun Square Juno in Male's Chart: Feels that wife is trying to hen-peck him or discount his masculinity. Building of lasting self-confidence better than upholding false values. Does not need false flattery, but true praise.

Sun Opposition Juno in Male's Chart: Difficult to adapt to softer or more feminine concepts. Enjoys an evening out with the boys or fishing trips.

Venus Square Juno in Male's Chart: Decidedly prefers secretaries, sisters, friends, etc., to wife concept. Usually tries to remake or change personality traits of the loved one after marriage, to change the mate from the type of woman he admires to one he can criticize. Nags a lot.

Venus Opposition Juno in Male's Chart: May at times be upset by money spent for artistic objects or home decoration. Emotional sexual frustration or frigidity which does not always manifest.

People often will marry and divorce time after time, not realizing that the basic fault is in their own personality rather than a mismatch with the partner.

The action of square aspects in the horoscope merely shows that the opportunity to learn certain lessons will be presented either through self or others. On the other hand, the opposition relates the need to balance the energies involved. As with a see-saw, the fulcrum or balance point is seldom exactly in the middle, but may be 60-40, or even 70-30. So rather than concern oneself with 50-50 relationships, it is better to aim toward true balance regardless of the ratio of give and take.

Remember that Juno deals only with the legalized relationship regardless of romantic love attachments. Juno deals with perpetuation of past traditions when marriage was considered more a business relationship rather than a love bond. Compatibility and love are not necessarily identical.

In addition to the consideration of the above-mentioned aspects, natal Juno by sign indicates the type of mate most often chosen by the native. The sign given need not be the Sun sign of the mate but what the mate expresses, her Moon or Venus Sign or his Sun, Saturn or Mars sign.

Following are interpretations for Juno in the natal chart:

Juno in Aries women look for strong, aggressive husbands who surge ahead into new adventures. These men like to be authority figures who judge other people. Often active in public affairs, these men can be irritable because they become involved in more projects than they can complete.

Juno in Aries men admire women who are actively involved. They prefer their wives to be either known for career or for community achievements. These may be women who privately have literal temper tantrums to get their own way.

Juno in Taurus women want husbands who are reliable and practical but who appreciate valuable possessions. They are looking for men who have strongly sensual natures as well as healthy bodies. These women admire men who are treasurers of large organizations or corporations. They are content with mates who enjoy farming for either profession or hobby.

Juno in Taurus men prefer wives who appear sweet and solid but are really iron fists in the velvet glove. They need the calmness of a soothing touch as much as the presence of live green plants in the home. These men do not object to their wives being status seekers as long as they remain cautious managers of family funds.

Juno in Gemini women want husbands who are playmates. (Surprisingly enough, Juno has a kindred relationship with Gemini because she was Jupiter's sister as well as his wife.) These women must have spouses who talk with them. Preferably they study new subjects and ideas together as well as travel extensively.

Juno in Gemini men prefer women they can father. They may be more advanced socially or more highly educated than their wives, thus making them like dependent children. These men like women who have a youthful appearance, windblown hairstyles and use very lightly scented perfume, if any. Legal alliances seem to hamper the sense of freedom of these men.

Juno in Cancer women seem happiest with husbands who fit the traditional patterns for a mate. They need men to willingly accept the responsibilities of providing for home and family. Compassionate, devoted husbands who spend most of their spare time around the house, especially in the kitchen or family room, would fill their needs for security.

Juno in Cancer men will consider themselves luckiest when married to gourmet cooks who are also excellent homemakers and mothers. These men want wives to appear feminine and gentle at all times and may wish to give them frequent gifts of jewelry, particularly pearls. They will be pleased with mates who wear both perfume and tasteful cosmetics. They tend to marry women who are quite moody in private.

Juno in Leo women are searching for the perpetual lover. They need to be courted throughout marriage as though it were the first date. They love flair and showmanship. A single red rose with a romantic note is more important to them than diamonds. The physical appearance of the mate must be handsome and impressive.

Juno in Leo men admire women who consider their beauty before their homemaking skills. These men would prefer to see their wives handsomely coiffured and manicured than have gourmet dinners. They enjoy sharing the spotlight with their mates. Lovely lingerie is a wise investment for women married to these men.

Juno in Virgo women demand fairness according to their own codes of justice. They want mates who are clean and meticulously dressed. These women like husbands who are particular about the details of daily life. They prefer men who have dependable but unassuming jobs like accounting. They prefer private, not public, demonstrations of affection.

Juno in Virgo men prefer wives who dress demurely in public and who are prudent about their handling of joint finances. These men function best with well-organized women who have very practical ideas. Although even tempered, it does not mean that these men are not sexually aggressive.

Juno in Libra women seek harmony in relationships. Husbands who are intelligent and artistic add cheer to their lives. These women enjoy graceful dancing partners, well-dressed gentlemen and pleasant conversationalists more than sports-oriented men. Physical grace is important to them.

Juno in Libra men like women who decorate their homes beautifully. They also are often married to their business partners. They believe in equal partnerships where both live up to every letter of the law. These men prefer wives who dress tastefully and have clear-textured complexions.

Juno in Scorpio women find discretion important in choosing husbands. They are not so much concerned with verbal communication as with sensitivity of the inner nature. These women prefer strong, silent men and will respect their need for solitude.

Juno in Scorpio men need women who are sensually beautiful and self-reliant. They prefer not to share too many of their inner feelings or concepts even with their mate. There is usually a physical magnetism between these couples.

Juno in Sagittarius women want freedom for everyone. These women enjoy husbands who travel, either physically or mentally, into new cultures. They look for optimistic mavericks rather than men who offer comfort and security.

Juno in Sagittarius men can be married to a career (rather than a physical mate) or the priesthood. These men need wives who understand their need to explore and expand. They do not admire women who are jealous or petty gossips. They often choose tall, rather large-boned women who are good conversationalists.

Juno in Capricorn women delight in husbands who run large businesses or organizations. These women like well-dressed, goal-oriented men who spend very little time at home. They respect hard work an effort. These women appreciate gifts of diamonds.

Juno in Capricorn men often marry women who are older or appear more mature. They do not object to marrying widows or divorcees with ready-made families. These men prefer small-boned women who dress rather conservatively and who are gracious hostesses for business dinners and parties. Discretion is an important trait for their wives.

Juno in Aquarius women restrict equality and friendship in marriage. These women want husbands who are outspoken, unorthodox and completely free, all the while knowing that these traits do not remain once these men are placed in the mold of husband.

Juno in Aquarius men adore free and outspoken women who are forever crusading for humanitarian causes. There is tremendous intellectual stimulation in these relationships but very few of the traditions of marriage. Often one mate either adores the other or demands subservience from him or her.

Juno in Pisces women look for spiritual mates. These women understand men who are dedicated to their careers, such as doctors, and do not resent the long hours they are away from home and family. They are sympathetic with husbands who follow their dreams rather than strive for material success.

Juno in Pisces men prefer women who are quiet, compassionate and peaceful. They like wives who work behind the scenes to lead people into preferred paths of thinking or action. These men admire subtlety and finesse. They would not respect women who were outspoken or immodest, especially in public.

Chapter V

Comparisons

There are several excellent books on synastry, or the study of human relationships through astrology. However, to date only *Planets and Asteroids* by Esther V. Leinbach has considered the effect of the asteroids Ceres, Pallas Athena, Juno and Vesta in chart comparison. The keywords given in this chapter for the asteroids blended with principles for aspects and other planets are sufficient to interpret asteroid effects in comparison charts.

An easy way to aspect one chart with another is:

1. Erect a flat or Aries-rising chart using zero degrees of the signs of the zodiac on their natural house cusps.

2. Then place the planets of one natal horoscope into the inner ring of the new chart.

3. With another color of pen or pencil place the planets of the other radix into the outer circle. Also include the respective Midheavens and Ascendants. Color coding makes it easier to remember who does what to whom.

4. Now aspect all planets of Native A to all planets of Native B using no more than a five-degree orb. This will give their instinctive reactions and feelings toward each other.

The major aspects between asteroids and other planets in the charts of both Elizabeth Taylor and Richard Burton are shown in Charts 5 and 6 with the comparison shown on Chart 7. This analysis will deal only with the asteroid aspects between the two charts.

Because of the square between Elizabeth's Ceres and Richard's Pluto, her health is shown to have suffered because of his intensity for living. Her Ceres trine his Moon as well as the conjunction between her Pallas and his Juno indicate their work together, involvement with Neptune adding that it was in the film industry. Burton's Neptune conjunction his natal Ceres would indicate that he uses his illusions in a very practical way—acting. However, notice the exact opposition between their respective Junos denoting that here we have two very dominate people in conflict over who will be more successful professionally.

Table 4. Keywords for Aspects of Asteroids in Comparisons

Degree	Aspect	Orb	Keyword
0	Conjunction	5	Intensity
60	Sextile	5	Understanding, similarity, compatibility
90	Square	5	Tension, activity
120	Trine	5	Blending, aid, peace
150	Quincunx	5	Need for revision
180	Opposition	5	Balance necessary

Elizabeth's Juno conjunct Richard's Pluto indicates not only that theirs was a sensual marriage but that she would tend to structure his emotions. She would frown on any extramarital affairs. With his Saturn and Sun trine her Juno, he would be prone to lavish her with jewelry and show her off as a grand possession. This tendency would show its reaction at the opposition of Juno to his Jupiter, showing his dislike of her extravagance. Elizabeth looked up to Richard as a father figure as well as a husband, as related to Sat-

Table 5. Keywords for the Asteroids in Comparison

Ceres	Pallas Athena	Juno	Vesta
Austerity	Agile	Bossy	Barren
Bandage	Allocate	Ceremony	Chaste
Care	Connect	Coiffure	Dedicate
Concern	Counsel	Eccentric	Discriminate
Cultivate	Defend	Faithful	Fired
Domesticate	Discipliner	Fruitful	Helpful
Farm	Earnest	Graceful	Honest
Feed	Guardian	Jealous	Perpetuate
Growth	Intuitive	Jeweled	Preserve
Harvest	Inventive	Married	Purity
Instinct	Liaison	Mature	Ritual
Initiator of life	Mathematical	Proper	Sacrifice
Nurse	Modest	Quarrelsome	Service
Nourishment	Organized	Rejuvenate	Solemn
Nurture	Pattern	Restrictive	Virginal
Practical	Peacemaker	Shrewd	Warm
Productive	Perceptive	Social	Zealous
Serve	Practical wisdom	Structures	
Useful	Protects	Subtle	
	Shielding	Veiled	
	Stern	Virtuous	
	Tamer	Uptight	
	Valiant		
	Works		

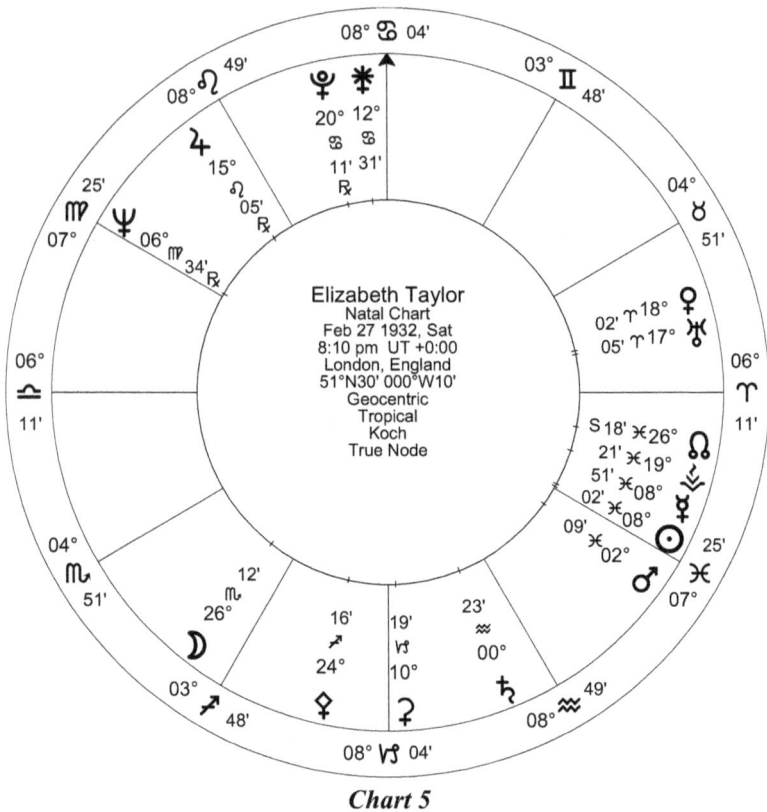

Chart 5

urn-Sun trine Juno. Her position as his wife (Juno) quincunx his Mercury (children) meant that his time and relations with his offspring had to undergo a revision because of her.

Elizabeth's Vesta added to the grand water trine found in Richard's chart; it was also conjunction his Uranus. Their hard work and dedication to their art cannot be denied; yet they brought scandal into each other's lives and created many sacrifices in personal areas.

There is no doubt that her beauty was an asset to their joint careers (Pallas trine her natal Venus), but her emotionalism (Moon) hin-

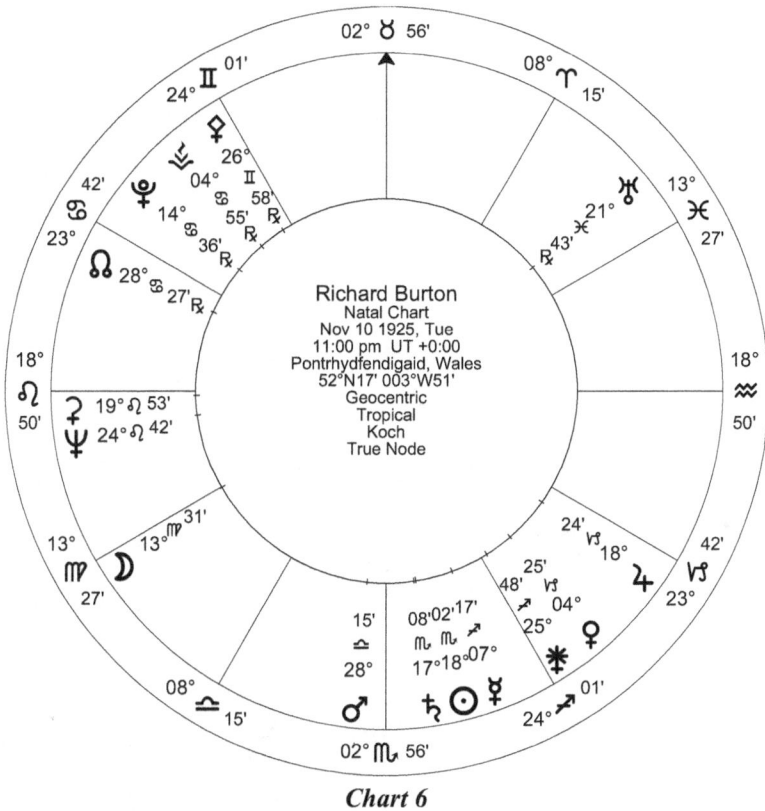

Chart 6

dered his career (quincunx Pallas). The Mars trine Pallas would mean that she energized the career or worked hard.

Richard's Vesta conjunction her Midheaven also shows the public scandals which were common during their married life. What was not shown publicly was Elizabeth's sensitivity (Neptune) and security conflicts (Saturn in aspect to his Vesta). His Vesta square her Ascendant shows she had difficulty coping with his career and goals taking precedence over her immediate demands.

Because the asteroids are so involved with daily living, it is difficult to know all of their connotations in public figures.

Inner Wheel
Elizabeth Taylor
Natal Chart
Feb 27 1932, Sat
8:10 pm UT +0:00
London, England
51°N30' 000°W10'
Geocentric
Tropical
0° Aries
True Node

Outer Wheel
Richard Burton
Natal Chart
Nov 10 1925, Tue
11:00 pm UT +0:00
Pontrhydfendigaid, Wales
52°N17' 003°W51'
Geocentric
Tropical
0° Aries
True Node

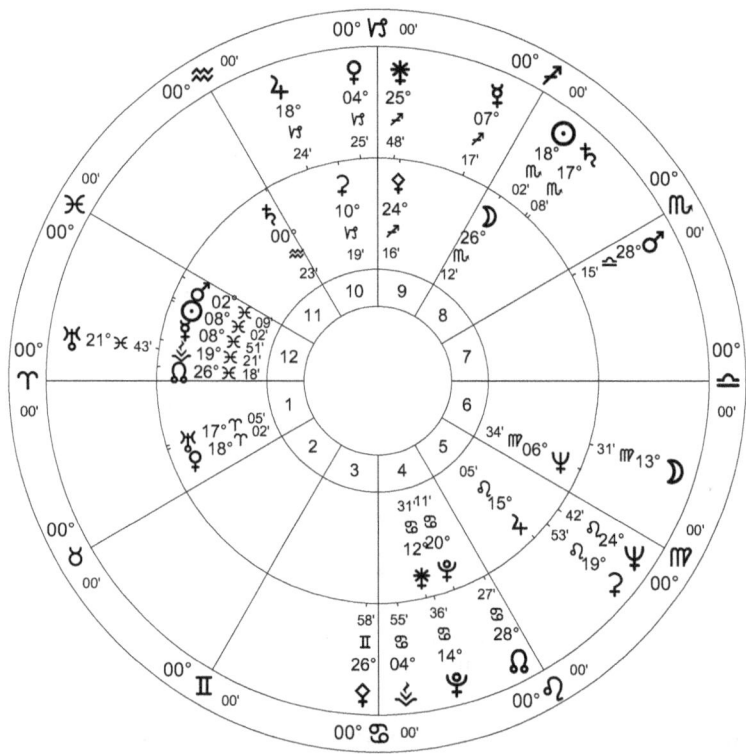

Chart 7. Aries Rising Comparison of Charts 5 and 6

Secondary progression positions of the asteroids were very active during 1974, the year of the Taylor-Burton final divorce. In Elizabeth's chart, progressed Ceres was opposition both natal Pluto and progressed Juno. Pallas had progressed to a square with progressed Mars and opposition to progressed Venus. Progressed Juno was conjunction progressed Pluto with progressed Vesta square natal Ceres and trine natal Pluto. Progressed Sun and Uranus were semisextile natal Vesta.

Richard's horoscope showed progressed Ceres conjunction natal Neptune as well as progressed Pallas and Mercury in exact opposition. Progressed Juno was trine natal Moon and opposition natal Pluto. Vesta had progressed to an opposition with natal Juno plus a sextile with natal Neptune. Progressed Venus was opposition natal Ceres.

In the comparison chart progressions, Elizabeth's progressed Ceres was quincunx his progressed Midheaven, while Richard's progressed Ceres was quincunx her progressed Ascendant. Elizabeth's progressed Pallas was opposition his natal Vesta and conjunction his natal Venus. Their progressed Junos were active with his semisextile her progressed Saturn and hers trine both his progressed Saturn and natal Uranus. Richard's progressed Vesta was opposition Elizabeth's natal Pallas in the year of their last separation.

As expected at the beginning or ending of a relationship, the asteroids were actively involved in exact aspects by progression. Since theirs was a business association as well as a marriage, Pallas was activated in addition to Juno and Ceres. There were more aspects in Elizabeth's natal chart, which explains why she initiated the divorce proceedings.

Interestingly enough, Elizabeth's first marriage to Richard occurred when her Sun had progressed to a square with natal Ceres. She was later married to John Warner when her progressed Sun was square her progressed Juno.

Chapter VI

Completion

When considering two natal or two progressed charts for relationship situations it is interesting to place the planets of each native's horoscope into the houses of the other's chart. To do this:

1. Place A's planets into B's natal chart houses or overlay.

2. Then place B's planets into A's natal chart houses. This same process may then be followed for their progressed planets into their respective progressed houses.

By using this method we see what attributes each person brings into the life of the other person. Many times there is the completion of configurations such as the grand trine, grand cross or yod. Elements lacking in one chart may be added by the partner, thus bringing balance to both.

While this method should never be used alone in evaluating the potential of a relationship, it does give additional insight.

On the following pages are principles or keywords for placement of the asteroids Ceres, Pallas Athena, Juno and Vesta of one person (A) into the houses of another person (B). Principles for this with the traditional planets may be found in the following books. *Com-*

parisons by Clara Darr and *Astrology of Human Relationships* by Sakoian and Acker.

Ceres
A's Ceres in B's houses mean that A adds the following advantages or problems for B (or self):

First House: Overly protective. Nurses through accident or illness. Adopts. Makes native feel useful and practical. Feeds. Shares difficult period or problem. Concerned. Mothers.

Second House: Supplies raw materials for information through which native may profit. Gives practical monetary instructions.

Third House: Encourages public speaking. Makes possible or necessary a study of nutrition. Supportive of creative writing. Adds in-laws to enlarge close family ties or responsibilities. Confides.

Fourth House: Brings pets into the home. Decorates with natural materials and plants. Nurses native or native's parents. Introduces new dietary patterns. Cares for native gently.

Fifth House: Helps rear children. Parents or counsels with native in absence of biological parents. Child-parent rapport. Child care.

Sixth House: Nurses through illness. Protects. Adds to communal income. Assists and encourages career stability.

Seventh House: Adds friends by marriage. Provides comfort when needed. Shelters. Aids career advancement. Publicizes native's talents. Promotes romance.

Eighth House: Loves like a brother or sister. Makes esoteric or hidden knowledge useable. Comforts through loss or bereavement. Lends a shoulder to cry on.

Ninth House: Supports or finances advanced education. Shows an appreciation of nature. Lives by simple moral code and gives native example to follow.

Tenth House: Gives practical advice and information about career endeavors. Supplies raw materials or wholesale goods for business growth. Hires for job. Concerned about native's health. Tries to prevent tension and overwork.

Eleventh House: Introduces either much older or younger friends to native. Teaches native to enjoy a simpler type of entertainment. Supports fulfillment of deepest desires. Aids development of self-confidence. Makes native feel needed.

Twelfth House: Accepts occult concepts of native as normal ideas. Shares spiritual experiences. Introduces drug or dietary changes. Nurses while hospitalized. Gives vitamins or medicine.

Pallas Athena
A's Pallas in B's houses mean that A adds the following advantages or problems for B (or self):

First House: Business partners who agree about goals and approach. Brings career openings. Helps realize ambitions. Supports native's image of self.

Second House: Provides financial support for career. Financial partners. Gives material and spiritual security. Energizes.

Third House: Supports training and education. Encourages self-expression.

Fourth House: Competes professionally. Trespasses into native's territory. Provides security for family. Establishes social codes for both.

Fifth House: Organizes social life. Plans mutual parties and vacations. Encourages financial risk. Inspires. Gives sexual security.

Sixth House: Counsels about career opportunities and investments. Gives a practical example to follow. Adds to work load.

Seventh House: Partners in business who have opposing career drives and methods. Challenges native's ambitions and purpose. Confrontations over security needs. Adds detachment in arguments or disagreements.

Eighth House: Enhances natural charm by making native more confident. Works for mutual security. Consoles others.

Ninth House: Presents opportunity to handle and absorb foreign cultural values. Introduces expanded concept of career possibilities. Finances travel for native. Pays for vocational training.

Tenth House: Extends financial credit. Supports publicly. Working in the same overall career field cooperatively.

Eleventh House: Gives confidence to realize greatest desires through working with groups and friends. Creates practical humanitarian actions out of native's philosophical ideals. Friendly about working conditions.

Twelfth House—Directs emphasis into service-oriented career. Financially supports native's psychoanalysis or spiritual development. Encourages growth.

Juno
A's Juno in B's house means that A adds the following advantages or problems for B (or self):

First House: Restricts. Criticizes appearance and dress. Makes native feel sloppy. Inhibits openness.

Second House: Forces to complete financial agreements made impulsively. Acts honorably about monetary agreements. Demands equality or an eye for an eye, tooth for a tooth.

Third House: Restricts communication. Insists on proper form of teaching and writing. Demands family acceptance of relationship. Makes native feel tongue-tied and cautious about phrasing statements.

Fourth House: Brings form into the home. Shares decorating plans. Structures family patterns. Controls family budget.

Fifth House: Inhibits love relationships. Promotes marriage(s). Expresses strong desire for children or grandchildren. Prohibits freedom for native.

Sixth House: Prods partner toward career advancements. Nags. Exacts much effort but rewards results generously. Demands that native use time productively. Preaches principle of hard work.

Seventh House: Demands that native live up to legal commitments. Protects native's reputation. Keeps relationship on formal basis. Can be physically unresponsive.

Eighth House: Expects native to realize full potential. Goads. Reacts prudishly to native's sexual openness. Embarrasses native. Physically restrictive. Keeps a tight rein on mutual finances.

Ninth House: Intolerant of hypocrisy. Travels luxuriously and to famous places or stays home. Structures the native's travel plans.

Tenth House: Criticizes publicly. Observes proprieties of teacher-student, employer-employee, husband-wife, etc. Warns of dire results of foolish actions. Threatens scandal.

Eleventh House: Lets native serve. Eliminates freedom with friends. Prods. Inhibits free expression of personality. Jealously demands native's time. Emphasizes strict codes of behavior.

Twelfth House: Deceives in marital affairs. Undermines self-confidence by restricting inner development. Frowns upon unconventional modes of expression. Concerned for mental and physical health.

Vesta
A's Vesta in B's houses means that A adds the following advantages or problems for B (or self):

First House: Expresses dedication and devotion through the partner. Shares common goals and sacrifices. Appears to the public to have denied native of basic needs, however is truly bringing benefits and growth. Basic nurturing. Takes away that which is not absolutely necessary.

Second House: Shares finances or food in times of shortage. Denies extravagance or foolish spending. Gives needed shelter and protection. Squelches spending sprees.

Third House: Denies family acceptance of relationship. Comes between native and friends. Separative. Devotes time to early training or career preparation.

Fourth House: Aids dedication in spiritual development. Shares giving up for purpose. Denies or delays family expansion. Forces absences from family. Helps native to understand purpose of humanitarian sacrifices. Empathetic.

Fifth House: Approaches relationship seriously. Accepts delays graciously. Jealous. May abandon native for prior commitments. Denies children.

Sixth House: Gives native a focus for self-denial. Presents opportunity for service. Challenges native to choose between career and duty.

Seventh House: Enhances conservative public image. Denies complete fulfillment in marriage. Annoys. Gives native strong realization of responsibilities.

Eighth House: Denies fruition of plans. Accuses native of unwise use of funds. Shares community or public concern.

Ninth House: Breaks down fixed ideas. Suggests and then delays travel plans. Questions ideas. Squelches optimism. Worries about native. Shares religious concepts.

Tenth House: Joins native on unusual quests. Delays fruition of ambitions. Presents obstacles to physical, financial and emotional success. Holds back promotions. Makes native feel needed.

Eleventh House: Emphasizes latent, humanitarian qualities. Remains loyal in face of mass desertion. Causes concern and worry.

Twelfth House: Shares zealous commitment. Psychic connections. Energizes for spiritual development.

After analyzing the effect which native A has upon native B, then reverse the process and read the effect native B has upon native A. In any form of synastry it is always important to remember who acts upon whom.

The charts of former president Jimmy Carter and his wife Rosalynn are used as an example of completion.

After erecting both charts, place Rosalynn's planets into the house cusps of Jimmy's natal chart as shown in Chart 10. Thus, we have A's planets in B's houses.

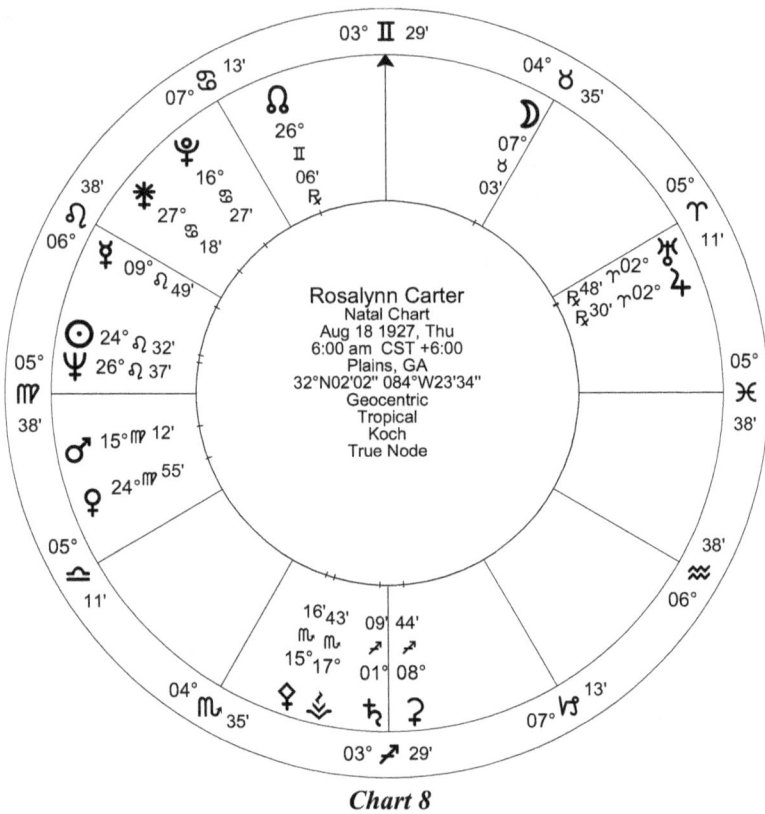

Chart 8

With Pallas in the first house they have been business partners in public. During the campaign she certainly provided career openings with her public appearances and speeches in his behalf. Everything would indicate that she shares his confidences and supports his image of himself.

Her trip to the Latin American nations fulfilled her position of Juno in his ninth house when she represented her mate and his country on a goodwill tour. She traveled in his stead.

With Vesta in his first house she willingly shares his sacrifices of time and family as well as his basic goals in life.

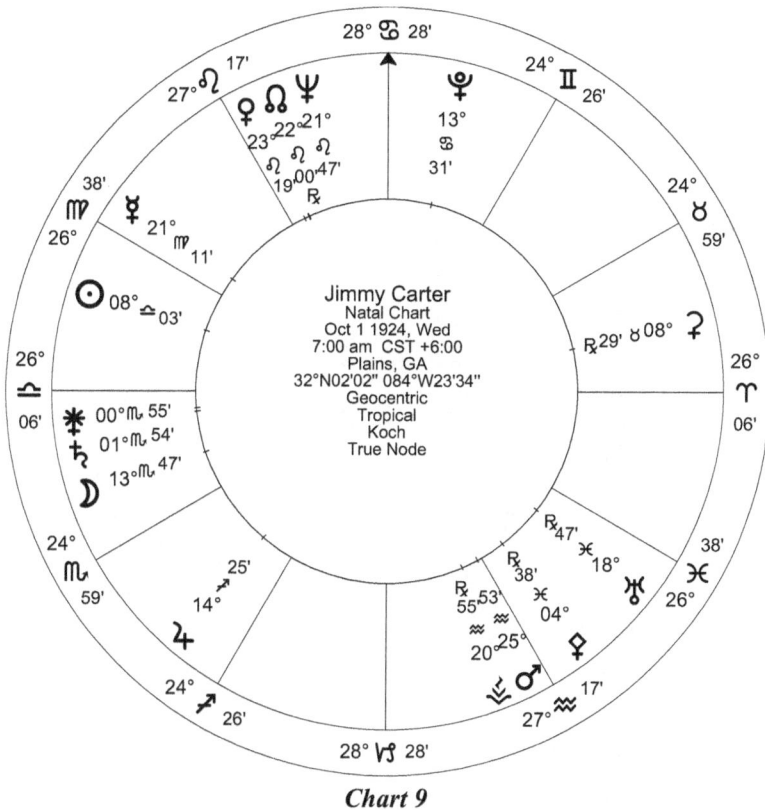

Chart 9

Ceres in the second house has been shown by her long-time service as bookkeeper and treasurer, not only of family funds but of the family business as well. If the advice she gave here had not been practical the business would surely not have been so successful. Certainly an enlarged analysis of the meaning of other planets in the chart would give additional information to this glimpse of the asteroids at work.

Secondly, place Jimmy's planets in Rosalynn's house cusps to learn how he influences her (Chart 11). Juno in the second house but on the third cusp will blend the values of both houses, indicating that he may be a demanding husband but that he provides hon-

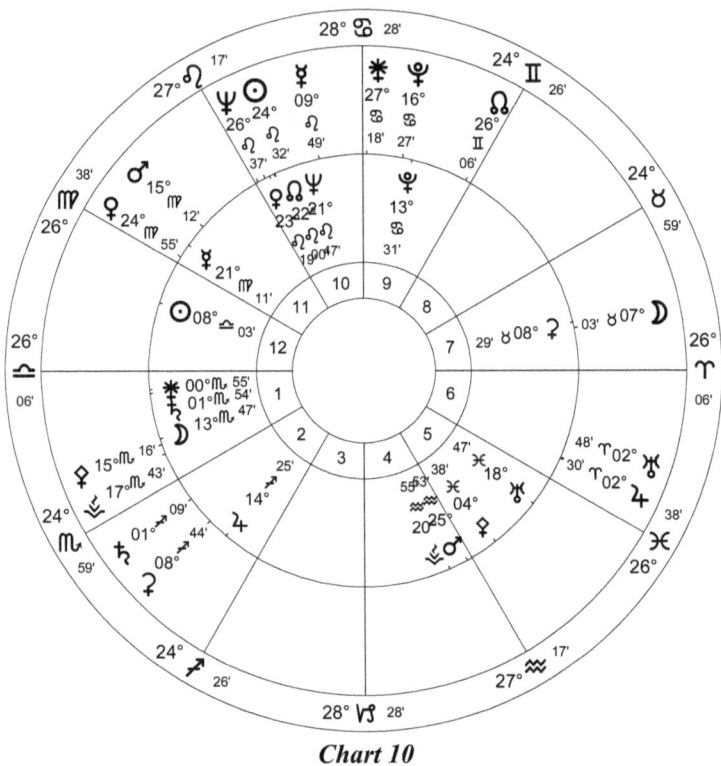

Chart 10

orably and fairly for his family. That his mate makes the most of her potential abilities is shown by the placement of Vesta in her sixth house.

With Ceres in Rosalynn's's ninth house, Jimmy has wholeheartedly supported his mate's foreign travel and educational endeavors.

Inner Wheel
Rosalynn Carter
Natal Chart
Aug 18 1927, Thu
6:00 am CST +6:00
Plains, GA
32°N02'02" 084°W23'34"
Geocentric
Tropical
Koch
True Node

Outer Wheel
Jimmy Carter
Natal Chart
Oct 1 1924, Wed
7:00 am CST +6:00
Plains, GA
32°N02'02" 084°W23'34"
Geocentric
Tropical
Koch
True Node

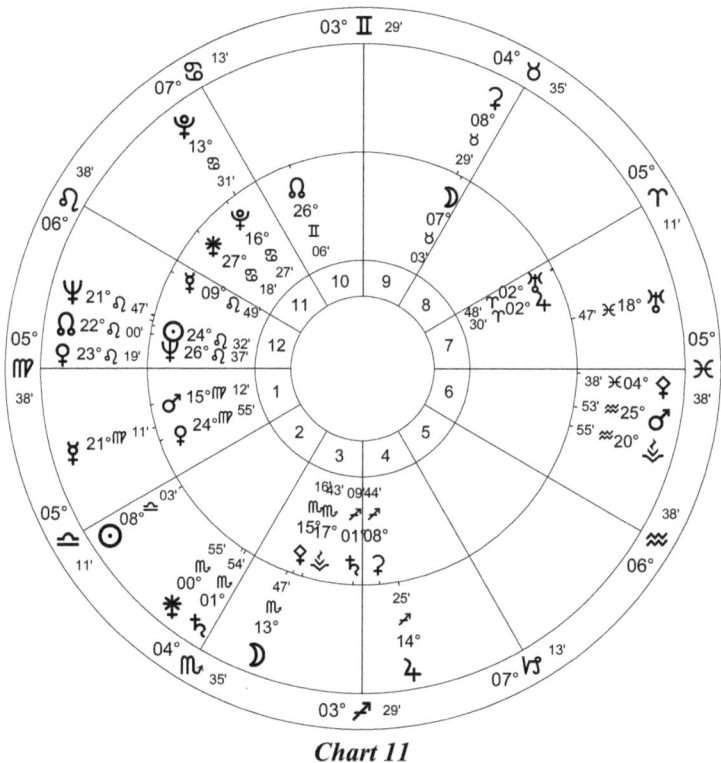

Chart 11

Pallas in exact conjunction with Rosalynn's seventh house cusp indicates again that they are partners in the truest sense, sharing their business and career advancements as well as family and friends. Many may remember his account prior to the election of their encounter over his decision to leave the navy to enter the family business. This would be the type of confrontation over security needs shown by Pallas in the partner's seventh house.

Chapter VII

Composite

Detailed information concerning the use of the midpoint or composite of planets in synastry may be found in other books. Suffice it here to briefly give a few basic rules for erection of the composite chart and use of the asteroids in this method.

1. Transpose all planets and points to the 360-degree wheel rather than use their sign positions.

2. Find the midpoint or half-sums of the Midheavens of two or more charts by adding together the values and dividing by the number of charts used.

3. Using this new composite Midheaven, erect a composite chart at the latitude where the relationship or group functions.

4. Now find the midpoints or half-sums of all planets plus the Ascendent.

5. The chart will have a locational Ascendant plus a composite Ascendant which is to be placed in the houses as if it were a planet called CA.

6. Aspect the new composite planets in the composite chart. See *Planets in Composite* for further analysis of the planetary meanings by placement and aspect.

7. To progress the composite chart, each natal chart must be pro-

gressed separately to the same day. Then recalculate the progressed composite chart from step 1.

8. Transits to the composite chart add much information about current happenings for the couple, family or group.

9. Some astrologers have successfully used midpoints and planetary pictures in delineating the composite charts.

By this midpoint system it is possible to see how the entire family unit or group reacts to various transits or events. It is interesting to see how relationships change at the birth of child, addition of a student into a classroom, substitution of members on a committee, use of a stand-in during a play performance, or any exchange of persons who are closely connected. Often when one person is rematched, such as in a divorce and remarriage, the public image seems to change. It is simply that the composite chart of the new marriage or relationship is quite different from the old. This makes the new pair appear as different persons, which they are not. Each individual will still react to his or her own natal horoscope as before.

Asteroids Ceres, Pallas Athena, Juno and Vesta have been found to have the following meanings in the houses of the composite charts of couples, families and groups.

First House—Public Image of the Group or Couple
Ceres: Very protective of each other. Nature-loving unless Ceres is badly aspected. Well fed. Not a biologically related group. Sometimes indicates foster or adopted children in a family unit.

Pallas: Business partners. Career-oriented. An ambitious couple. Materialistic family or group. Successful unless badly aspected.

Juno: Concerned with appearances. Very strict about living up to mutual agreements. Appears to achieve a 50-50 relationship but this takes its toll on the personal rapport.

Vesta: Charitable enterprise or couple. Zealously involved in some project or activity together. May belong to unusual religious sect which makes strict demands on its members.

Second House—Material Concerns and Goals
Ceres: Basically careful about handling money. Work for combined income. Receive income from services, work with animals, natural food store or medical profession.

Pallas: Financial support from job(s). Work for practical purposes and to acquire possessions. Purchases are more useful than aesthetic.

Juno: Concerned with finances. Overspend on luxury items (especially while traveling). Numerous arguments about money.

Vesta: Personal self-denial on the part of each native for the security of the group or marriage. Unorthodox attitudes about money. Charitable. Philanthropic.

Third House—Communication with Each Other
Ceres: Mental rapport but does not promise physical congeniality. A free flow of communication about everyday matters, especially children and food.

Pallas: Work together as secretary-boss or owner-employee rather than co-owners. Give financial aid to in-laws for such things as education. Discuss career plans rather than fun activities. Hobbies become vocational pursuits.

Juno: Communicate more as siblings than man-woman or parent-child. Formal and constrained relations with in-laws. Shy.

Vesta: Sacrifices because of children in a marriage. Long-distance courtship. Lack of communication on mundane matters in the relationship.

Fourth House—Home and Family
Ceres: Parental approach of one partner for the other. Much care and concern for the well-being of the total family. Healthy.

Pallas: Business enterprises in the home. All or both interested in advanced training or education.

Juno: Restrictive. Jealousy of one mate about the other's activities. Frigidity in marriage not manifested by either separately.

Vesta: Sadness or sacrifice in the home at some period during the relationship. Share a family burden of illness or bereavement.

Fifth House—Hobbies, Children and Affections
Ceres: Sexually compatible. Affectionate with each other even in public. Enjoy simple pleasures and activities.

Pallas: Gracious toward one another. Tendency to financial extravagance. All parties contributing funds from separate income or investments.

Juno: Restrictive physically. Concern with fitting in to the accepted social patterns. Not conducive to a love affair. React in a proper manner. Social climbers.

Vesta: Denial of children. Raising offspring from previous marriages. Not a smooth running love affair. Delays to happiness because of prior responsibilities of one or both parties.

Sixth House—Work, Diet and Pets
Ceres: Sharing different cultural backgrounds on a daily basis. Learn new patterns together. Nurse or tend one another. Lighten each other's load just by being around.

Pallas: Hard work together building a business. Career interests demand serving and supporting each other.

Juno: Working couple concerned with separate careers. Conflict of hours and interests may make one or the other feel ignored.

Vesta: Conflicting demands on time so there is very little personal life. Share marital or group responsibilities equally and willingly.

Seventh House—Cooperation
Ceres: Outer harmony of bodies and temperaments but inner tension. Some conflict with daily habits. Different diets.

Pallas: Cooperation in mental activities. Work together. Different organizational techniques. Family business. Financial support for career training shared.

Juno: Both see realization of goals as permanent togetherness. Refusal to compromise. Demand strict rules of behavior from each other. Unforgiving after argument.

Vesta: Receive social reproval for not abandoning an unorthodox member of the family or group. All suffer because of denial of conventional behavior. Unusual family or unit.

Eighth House, Joint and Family Possessions, Sex
Ceres: Siblings. Good physical rapport. Concern with the finances or health of partner's parents.

Pallas: Pay for psychoanalysis for a member of the group. Support unusual sexual liaisons. Work for in-laws.

Juno: Strife between husband and wife. Tendency to secret or unacceptable love relationships on the part of one or more of the group or family members.

Vesta: Periods of celibacy or sexual perversion. Care of relatives without remunerations. Will monies to charitable or religious organization.

Ninth House—Philosophy and Religion
Ceres: Groups with varied ethnic or racial backgrounds. Harmonious blending of cultures. Children have mixed genetic heritage.

Pallas: Travel together for business purposes. One funds the partner's travels. Paid members of the church or synagogue choir. Co-workers in religious organization. Religious vocation.

Juno: Mixed marriage. Travel restricted for lack of funds or else tour like royalty. Conservative about religious involvements.

Vesta: Different philosophical approaches. Over-zealousness on part of one hinders the growth of other party (parties). Unwise actions because of illogical dedication to a theory or group.

Tenth House—Reputation and Ambition
Ceres: Cautious about public reactions. More supportive than sexual. Friends. Parental attitude toward one another. Gentle appearance. Natural grace.

Pallas: Both have professional training for their careers. Desire for monetary success and power. Materially supportive. Can be competitive if both parties work together.

Juno: Uptight about reputation as a couple. Appear aloof. Careful adherence to protocol. Dresses rather formally in all circumstances. Expensive tastes.

Vesta: Delays or denials of success together. Creates public disfavor or scandals. Shows extreme dedication to cause or project to the exclusion of another's interests.

Eleventh House—Mutual Friends
Ceres: Investigates natural foods and remedies with friends. Enjoys entertaining at impromptu parties. Nurses friends through illness.

Pallas: Works with social acquaintances. Success at career for one or both parties. Social work.

Juno: Adjustment of goals to mutual satisfaction. Restrictive about casual friendships. Enjoyments rather expensive and formal. Distant.

Vesta: Delay of fulfillment of wishes. Friends ignored because of family problems or responsibilities. Age difference in couple creates lack of compatible interests.

Twelfth House—Hidden Feelings and Motives, Illnesses
Ceres: Poor health. One partner or family member nurses another through long illness or hospitalization. Supportive through drug cure.

Pallas: Work in medical institution. Pay for extensive health problem. Study first aid together.

Juno: Marriage may ruin harmony of this relationship. Feel trapped by rules and laws. Restrictive to personal freedom. Very sensitive individuals who have trouble blending.

Vesta: Dedicated to spiritual and hidden goals. Must be careful not to harm each other out of over-zealousness. Self martyrdom. Excessive sacrifices may be undoing of partnership.

The young couple whom we shall use as the example for the composite method were interested in two aspects of their chart which are particularly influenced by the asteroids: forming a business partnership and deciding about a permanent personal liaison.

Charts 12 and 13 were midpointed to determine the placements for Chart 14, or the composite chart. Then the composite planets were aspected as though in a natal chart.

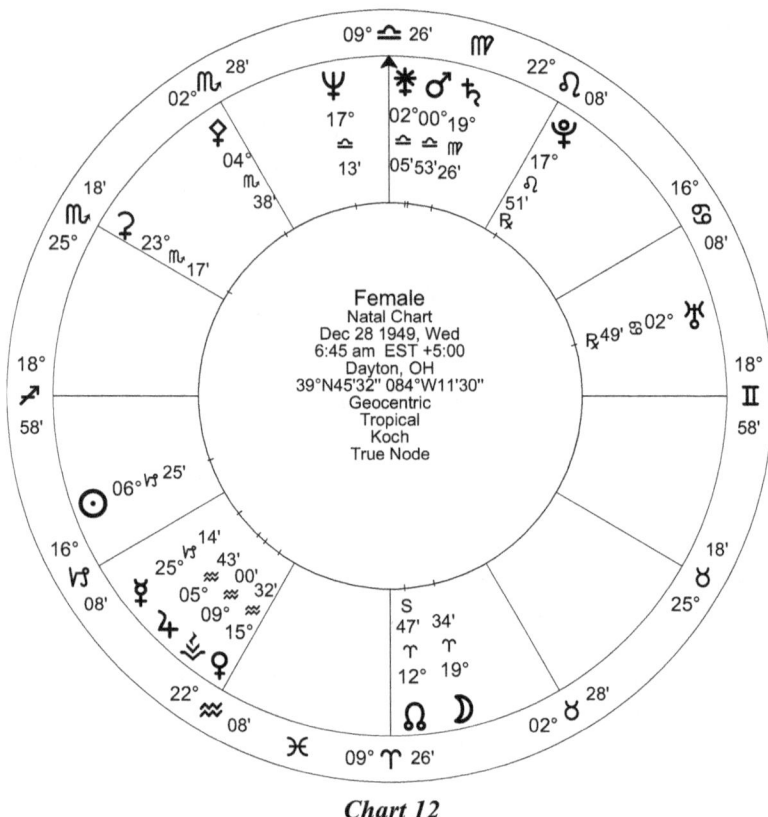

Chart 12

Considering only the aspects to and from the asteroids, a T-square is found involving Pallas, Pluto and Jupiter, indicating potential conflict between the two concerning home, power and career. He wants a more traditional type of wife (see his natal Juno in Libra) and she needs a great deal of freedom and equality in relationships. However, a trine from Pallas to Vesta will help this situation by giving dedication and hard work to the business endeavor. It also suggests that there will not be the extra complication of having children together. Applying semisextiles from Pallas to both Neptune and Juno gives their idealism and practicality a chance to emerge.

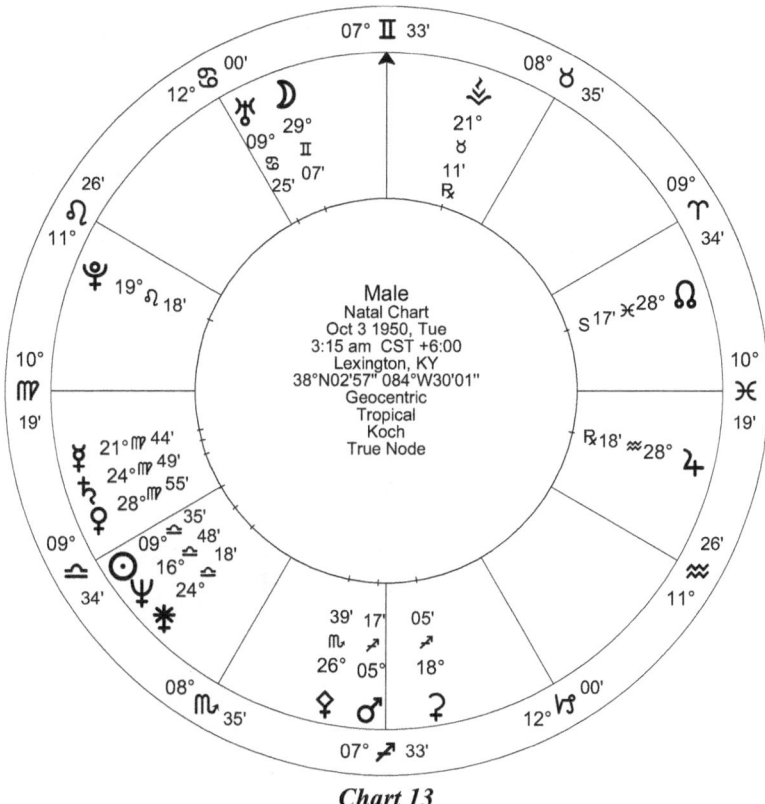

Chart 13

Ceres is trine the North Node of the composite chart, indicating that the public will like and respect them. Also, Juno is sextile Ceres, giving this couple a good chance to earn money by adherence to proper business principles. Though they seem involved in new concepts and ideas, they are both cautious and considerate. With Ceres in the second house it was interesting to learn that they were contemplating buying a pet shop.

The fifth house Vesta is quincunx both Pluto and Neptune, forming a yod, or finger of fate. Continuing dilemmas in the career will force constant revision of the personal or inner values for these two young people.

Twelfth house is not the best position for planets concerned with relationships, but their Juno in the twelfth house could be used as inner or spiritual growth together rather than the self martyrdom usually indicated on the part of one partner or the other. Juno trine Jupiter in the fourth gives them a lovely and expensive home, but opposition the North Node and quincunx Vesta presents some problems concerning their friends and the lack of children.

Chapter VIII

Contact Cosmogram

For those astrologers who are accustomed to using cosmobiology or the Uranian 90-degree wheel, the cosmic cosmogram is simple. For others, it would be necessary to examine the method more extensively by reading literature by Ebertin, Witte or Kimmel on this method of astrology.

1. Take two erected horoscopes which have been placed on 90-degree wheels.

2. Place Chart A on the inner ring of the contact cosmogram and place Chart B on the outer ring of the contact cosmogram.

3. Then read midpoints and planetary pictures as usual, considering points from both horoscopes rather than one alone.

4. Direct in the usual manner using separate directions for both persons involved. For two persons this would fill four circles of the contact cosmogram.

5. Consider transits also.

Placing all horoscopes involved on the 90-degree wheel automatically shows conjunctions, squares, semisquares, and sesquiquadrates thereby describing tension and activity between the couple or in the group. There is also the potential for seeing whether either person has planets aspecting midpoints in the other's chart.

General keywords for the asteroids may be used for delineation with midpoints. Ceres brings the concept of nurturing wisely and well without becoming personally involved. Pallas Athena trains, disciplines, defends and protects all toward the goal of cooperative productivity. Moody Juno can be either warm and gentle or cold and critical according to aspect. Vesta brings willing sacrifices in areas of commitment. See also the earlier keyword pages for each asteroid.

In one example, a woman's Juno/Vesta was conjunct her mother's Saturn. True to expectations, this native was denied marriage or close liaisons because of caring for her aged mother.

Where a man's Saturn/Uranus was touched by his wife's Pallas there were unusual marital problems brought about by strange business hours. He worked days and she worked nights!

Fluctuating joint finances did truly create tensions for a couple where the husband's Mars conjunct his wife's Juno/Jupiter.

The use of synastry is often restricted to a discussion of marital problems or love relationships. However, it is equally valid in considering all situations where people interact. In the following example, the contact cosmogram will be used to show how four well-known United States senators automatically react to each other.

A brief look at the contact cosmogram shows all of the squares, conjunctions and oppositions. Senator Humphrey's Juno aspects Senator Percy's Pluto, thus giving him cause to make long-term revisions in his welfare concepts. His Vesta is also on Senator Jackson's Juno, making them both zealous workers but rather set in their opinions. Senator Jackson has Pallas aspecting Senator Percy's Venus and Senator Goldwater's Neptune, Ascendant,

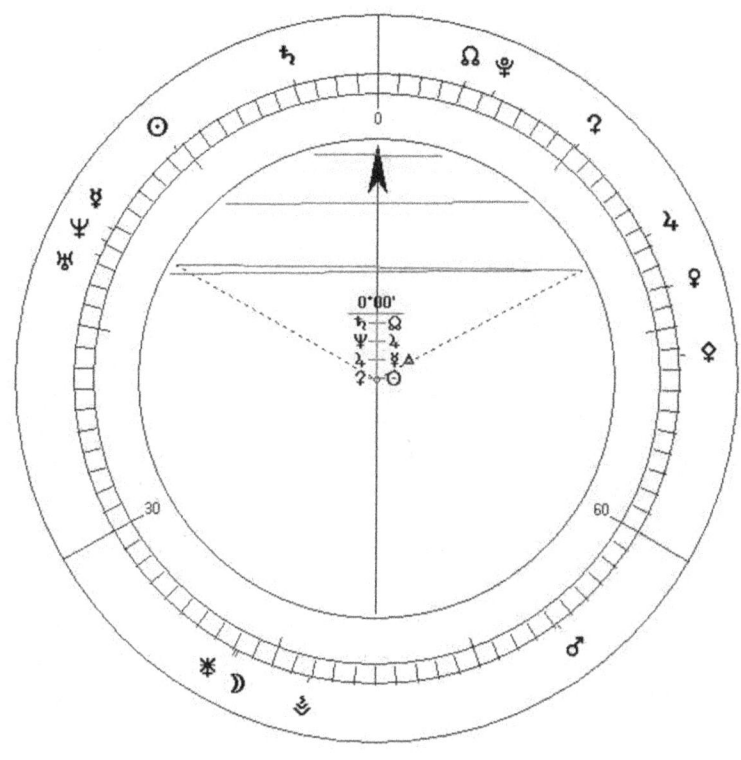

Chart 15. Barry Goldwater

Uranus and Midheaven, showing that he is closer in concept to Charles Percy than Barry Goldwater.

Senator Percy's Pallas aspects Senator Humphrey's Mercury and Goldwater's Vesta. Jackson's concern for others would be zealously opposed by Senator Goldwater while vocally supported by Senator Humphrey. Senator Percy's Ceres on Senator Humphrey's Pluto shows admiration and concern for a fellow congressman even though they have opposing views.

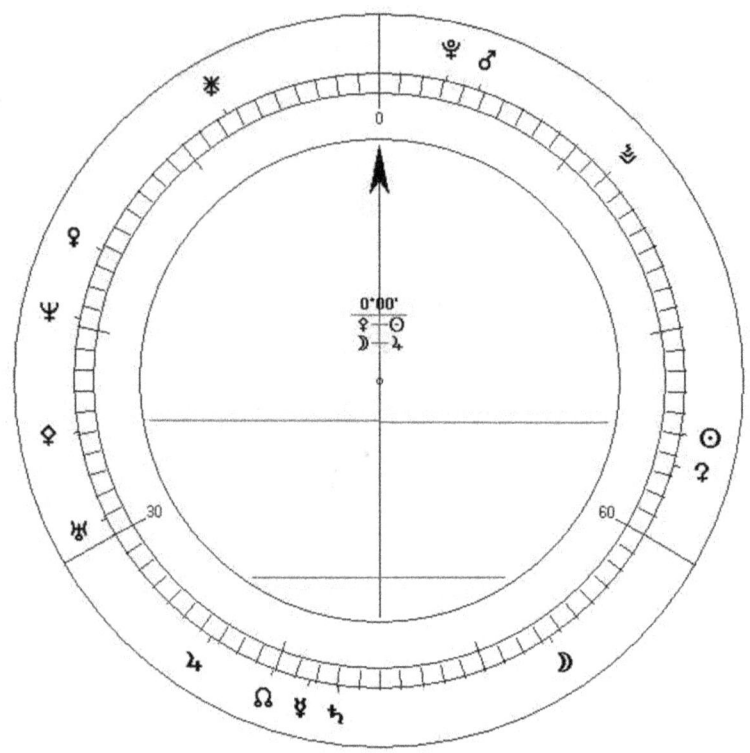

Chart 16. Hubert Humphrey

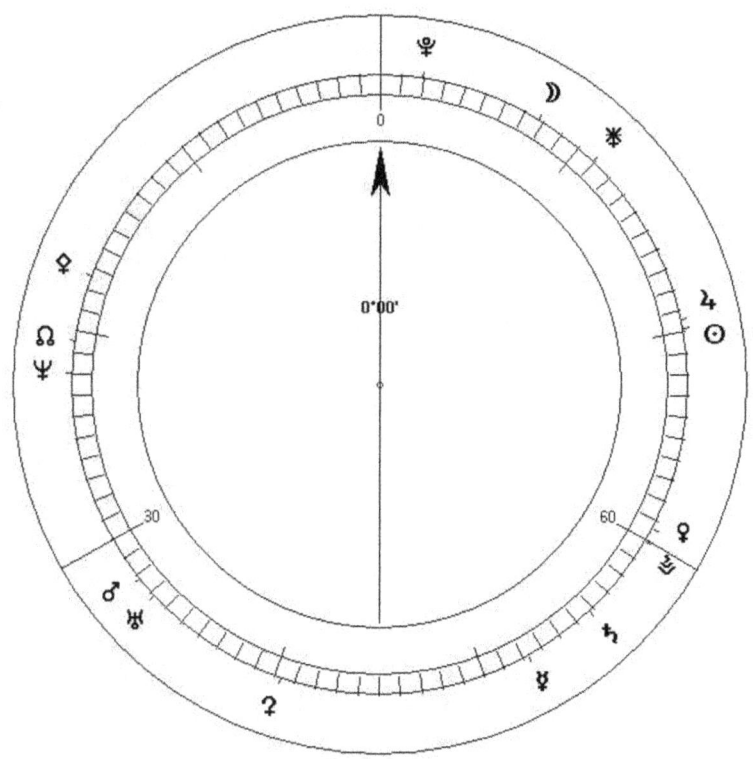

Chart 17. Henry Jackson

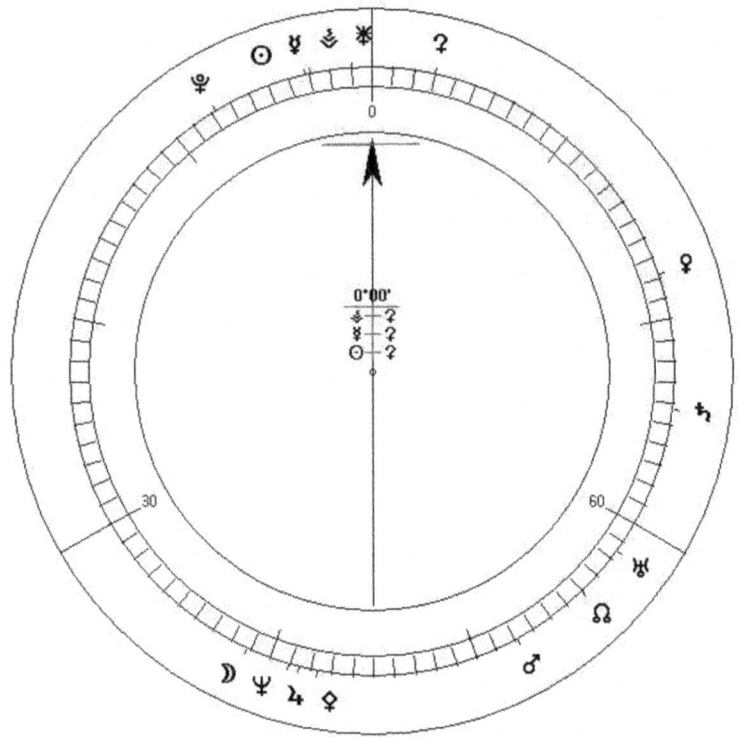

Chart 18. Charles Percy

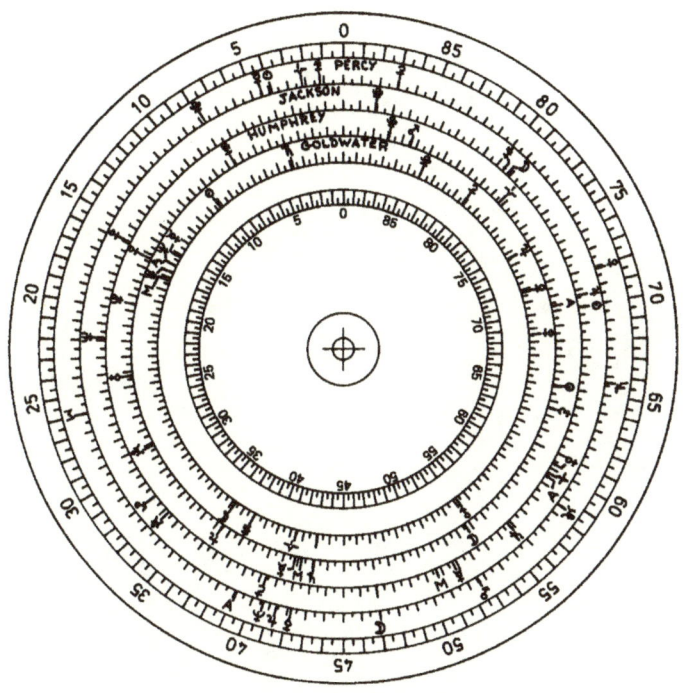

Chart 19. Contact Cosmogram of Four Senators

www.ingramcontent.com/pod-product-compliance
Lightning Source LLC
Chambersburg PA
CBHW051710040426
42446CB00008B/806